INTEGRATED RESOURCE AND ENVIRONMENTAL MANAGEMENT

Integrated Resource and Environmental Management
Concepts and Practice

Edited by
Kevin S. Hanna
D. Scott Slocombe

University Press

OXFORD
UNIVERSITY PRESS

70 Wynford Drive, Don Mills, Ontario M3C 1J9
www.oup.com/ca

Oxford University Press is a department of the University of Oxford.
It furthers the University's objective of excellence in research, scholarship,
and education by publishing worldwide in

Oxford New York
Auckland Cape Town Dar es Salaam Hong Kong Karachi
Kuala Lumpur Madrid Melbourne Mexico City Nairobi
New Delhi Shanghai Taipei Toronto

With offices in
Argentina Austria Brazil Chile Czech Republic France Greece
Guatemala Hungary Italy Japan Poland Portugal Singapore
South Korea Switzerland Thailand Turkey Ukraine Vietnam

Oxford is a trade mark of Oxford University Press
in the UK and in certain other countries

Published in Canada
by Oxford University Press

Copyright © Oxford University Press Canada 2007

The moral rights of the author have been asserted

Database right Oxford University Press (maker)

First published 2007

All rights reserved. No part of this publication may be reproduced, stored in a retrieval system, or transmitted, in any form or by any means, without the prior permission in writing of Oxford University Press, or as expressly permitted by law, or under terms agreed with the appropriate reprographics rights organization. Enquiries concerning reproduction outside the scope of the above should be sent to the Rights Department, Oxford University Press, at the address above.

You must not circulate this book in any other binding or cover
and you must impose this same condition on any acquirer.

Library and Archives Canada Cataloguing in Publication

Integrated resource and environmental management : concepts
and practice / edited by Kevin S. Hanna and D. Scott Slocombe

Includes biographical references and index.
ISBN 978-0-19-542049-4

1. Natural resources–Management–Textbooks. 2. Environmental management–Textbooks. I. Hanna, Kevin S. (Kevin Stuart), 1961– II. Slocombe, D. Scott (Donald Scott), 1961–

GE300.I582 2007   333.7   C2007-901190-X

Cover image: Ron Niebrugge
Cover design: Sherill Chapman

1 2 3 4 – 10 09 08 07
This book is printed on permanent (acid-free) paper ∞.
Printed in Canada

Contents

TABLES AND FIGURES		vii
CONTRIBUTORS		ix
PREFACE		xii

1   Integration in Resource and Environmental Management   1
D. Scott Slocombe and Kevin S. Hanna

2   An Overview of Integration in Resource
and Environmental Management   21
Bruce Mitchell and Dan Shrubsole

3   Research and the Integration Imperative   36
Stephen Dovers and Richard Price

4   Governance for Integrated Resource Management   56
Ann Dale and Lenore Newman

5   Integration through Sustainability Assessment: Emerging
Possibilities at the Leading Edge of Environmental Assessment   72
Robert B. Gibson

6   Integrated, Adaptive Watershed Management   97
Bruce P. Hooper and Chris Lant

7   Implementation in a Complex Setting: Integrated
Environmental Planning in the Fraser River Estuary   119
Kevin S. Hanna

8   Resolving Human–Grizzly Bear Conflict: An Integrated
Approach in the Common Interest   137
Seth M. Wilson and Susan G. Clark

9   Collaborating Experts: Integrating Civil and Conventional
Science to Inform Management of Salal (*Gaulthenia shallon*)   164
Heidi Ballard and Louise Fortmann

| 10 | Integrating Scientific Information, Stakeholder Interests, and Political Concerns | 181 |

Lawrence Susskind, Patrick Field, Mieke van der Wansem, and Jennifer Peyser

| 11 | Dis-integrating Equity: Sustained-Yield Forestry and Sustainability in Vallecitos, New Mexico | 204 |

Carl Wilmsen

| 12 | Information Management for Water Resources: Concepts and Practice | 220 |

Sarah Michaels, Daniel McCarthy, and Nancy P. Goucher

| 13 | Integrated Approaches for Transboundary Wildlife Management: Principles and Practice | 236 |

Michael S. Quinn

| Conclusion: Reflections and Prospects for irem | 252 |
| Index | 256 |

Tables and Figures

TABLES

1.1 Some commonly identified operational factors and elements for successful integrated resource and environmental management
4.1 Institutional characteristics that support sustainable development
5.1 The structure and key components of potentially effective assessment processes
5.2 General sustainability requirements as criteria for sustainability assessment decision making
5.3 The basic design features of best-practice sustainability assessment processes
5.4 Possible general rules for decisions about trade-offs and compromises
6.1 Institutional, organizational, and human resource elements of integrated, adaptive watershed management
8.1 Comparison of Rocky Mountain Front and Blackfoot Valley case studies, Montana, USA
12.1 Comparing information management and first- and second-generation knowledge management
12.2 Components of conservation authorities' perspective on information management
12.3 Comparing steps in Choo's (1998) process model of information management with the components of conservation authorities' perspective on information management

FIGURES

4.1 Enlarging the context of government policy-making
6.1 A watershed as a complex human-environment adaptive system
6.2 An ecological-economic production possibilities frontier and social preferences indifference curves
6.3 An approach to watershed modelling to facilitate integrated, adaptive management within a political context
6.4 The hypothesized qualitative relationship between net carbon flux and other economic and ecosystem service goals
7.1 Area of the Fraser River Estuary Management Program
7.2 Fraser River estuary shoreline near Point Grey
7.3 Log booms at Iona Island
7.4 Fraser River Estuary Management Program implementation structure
8.1 Location of Rocky Mountain Front and Blackfoot Valley case studies in Montana, USA

8.2 Number of livestock carcasses removed from project area in Blackfoot Valley, Montana, during 20 February–15 May 2003–2006
9.1 Salal and bracken fern form a dense understory in intensively managed stand of Douglas fir
9.2 A salal harvester scientist records information on data sheets translated into Spanish
10.1 The consensus-building process and the role of joint fact finding
10.2 Joint fact finding: Key steps in the process
11.1 Location of the Vallecitos Federal Sustained Yield Unit
11.2 The VFSYU and surrounding communities
13.1 Satellite locations for female wolf 'Pluie', originally collared in Peter Lougheed Provincial Park, Alberta

Contributors

HEIDI BALLARD
Assistant Professor of Environmental Science Education
School of Education
University of California, Davis

SUSAN W. CLARK
Professor, Yale School of Forestry and Environmental Studies and
 Northern Rockies Conservation Cooperative

ANN DALE
Trudeau Fellow and Professor, Science, Technology and Environment Division
Canada Research Chair in Sustainable Community Development
Royal Roads University

STEPHEN DOVERS
Senior Fellow, Centre for Resource and Environmental Studies
Australian National University

PATRICK FIELD
Managing Director, Consensus Building Institute
Cambridge, Massachusetts

LOUISE FORTMANN
Professor, Department of Environmental Science, Policy and Management
University of California, Berkeley

ROBERT B. GIBSON
Professor, Department of Environment and Resource Studies
University of Waterloo

NANCY P. GOUCHER
Resource Knowledge Group, School of Planning
University of Waterloo

KEVIN S. HANNA
Associate Professor, Department of Geography and Environmental Studies
Wilfrid Laurier University

BRUCE P. HOOPER
Principal, River Basin Management
DHI Water and Environment, Australia

CHRIS LANT
Professor, Department of Geography
Southern Illinois University

DANIEL MCCARTHY
Resource Knowledge Group, School of Planning
University of Waterloo

SARAH MICHAELS
Associate Professor, Resource Knowledge Group, School of Planning
University of Waterloo

BRUCE MITCHELL
Professor of Geography and Associate Provost
University of Waterloo

LENORE NEWMAN
Assistant Professor, Science, Technology and Environment Division
Royal Roads University

JENNIFER PEYSER
Facilitator, Resolve
Washington, DC

RICHARD PRICE
Director, Kiri-Ganai Research
Canberra

MICHAEL S. QUINN
Assistant Professor, Faculty of Environmental Design
University of Calgary

DAN SHRUBSOLE
Associate Professor, Department of Geography
University of Western Ontario

D. SCOTT SLOCOMBE
Professor, Geography and Environmental Studies
Wilfrid Laurier University

LAWRENCE SUSSKIND
Ford Professor of Urban and Environmental Planning
Head, Environmental Policy Group
Director, MIT-Harvard Public Disputes Program
Department of Urban Studies and Planning
Massachusetts Institute of Technology

MIEKE VAN DER WANSEM
Director, Consensus Building Institute
Cambridge, MA

CARL WILMSEN
Coordinator, US Community Forestry Research Fellowship Program
College of Natural Resources
University of California, Berkely

SETH M. WILSON
Research Affiliate, Yale School of Forestry and Environmental Studies
Adjunct Faculty, Environmental Studies Program, University of Montana, and Northern Rockies Conservation Cooperative

Preface

Over the last 30 years and more, resource and environmental management (REM) has gone through a series of manifestations, from multiple use to integrated watershed management, integrated resource management (IRM), comprehensive regional land-use planning, and ecosystem-based management. Common to all these is an interest in integration, which at minimum has several meanings—integration across disciplines, integration across agencies, and integration across sectors. Whatever label it appears under, integration has also emerged as a prominent theme in the policy and management rhetoric of virtually every resource and environmental management agency in North America and abroad. More specifically, and more recently, there has been great interest in integration from the perspectives of particular resources and approaches, such as watersheds and forest resources, and growing attention to integration in planning from a scientific, informational, and multiple-values perspective. Integration is not only an integral factor in resource and environmental management, but as the chapters of this book show, it is also an increasingly complex notion, one that draws from a range of interpretations and approaches.

While many works have addressed integration as a component of resource and environmental management, very few have done so systematically or have had REM as the primary focus. But there have been exceptions. R. Lang's 1986 edited volume, *Integrated Approaches to Resource Planning and Management*, provided a collection of case studies based on explorations of integrated resource management; Bruce Mitchell's comprehensive 1979 work, *Geography and Resource Analysis*, examined the spatial dynamics of a diverse range of natural resource management issues; and W.A. Duerr and colleagues' 1982 *Forest Resource Management*, while it certainly centred on forestry practice, also included notably broad integrative elements geared towards incorporating diverse forest uses into decision making. Although all of these books are still worth a look, they are difficult to find and much has changed from both theoretical and case study perspectives since they were published.

This book is inspired by such approaches, and our purpose is to update and expand the notion of integration, building on the substantial practical experience and theoretical development in resource and environmental management over the last 30 years. In the following chapters we develop the concept of IRM beyond discussions of specific resources and work towards an integrative framework based on a focused, consistent discussion of issues and challenges, supported by specific case studies and illustrations of approaches and methods. This volume thus includes critical coverage of existing approaches and new opportunities.

This collection of specially commissioned papers provides a comprehensive, coherent, and multidisciplinary anthology with a diverse range of examples from across North America and elsewhere. We anticipate that it will be of interest to academics and resource and environmental professionals, as well as being especially helpful to students, whether at the undergraduate or graduate level, in geography, planning, resource management, forestry, and environmental studies programs—indeed in all realms of resource and environmental management.

While the earlier chapters tend to take a historical and conceptual approach, most also contain case studies. In chapter 1, Slocombe and Hanna discuss the history of integration broadly, emphasizing resource and protected-area management, and examine the dimensions, challenges, and opportunities in integration for resource and environmental management. In chapter 2, Mitchell and Shrubsole also address the history of integration, but more briefly and with an emphasis on forestry, watershed management and development, and sustainable development, using case studies of the US Forest Service and Ontario conservation authorities to develop lessons and conclusions.

In chapter 3, Dovers and Price discuss types of integration, when to integrate and when not to, integration motives and interests, and dimensions of integration (notably why and how, as well as participation, scale, and expertise) before going on to discuss integration in research, policy, and management, looking at a Land and Water Australia case study and the national dryland salinity program; they conclude with lessons and guiding principles for integrated research. In chapter 4, Dale and Newman address governance and integration, emphasizing the goal of sustainable development; they consider such topics as demands for agency and participation by publics, the need for sustainable development, government structure problems in achieving new approaches and processes, the implementation gap, institutional characteristics for fostering sustainable development, and the need for enlarged policy development and decision-making contexts. In chapter 5, Gibson focuses on the evolution of environmental impact assessment as a tool for integration and particularly on sustainability assessment as the new cutting edge of environmental assessment and as a key integrative focus. Core topics include sustainability requirements as a basis for sustainability assessment, sustainability assessment process basics, and approaches to trade-off decisions.

Succeeding chapters have a strong sectoral, case study orientation, although several also include some theoretical background and lessons for theory and practice. In chapter 6, Hooper and Lant draw on systems and panarchy theories to explore managers' and landowners' views of watershed management in developed and developing countries, the challenges and issues in integration, the nature of adaptive management approaches (including benefits and disbenefits), and the role of incentive and property rights approaches in fostering integration in a watershed management context. Hanna, in chapter 7, looks at issues inherent in implementing an integrated approach and explores this element through a case study of British Columbia's Fraser River Estuary Management Program, highlighting the importance of the implementation process to the success of integrated approaches. His chapter also provides a framework for characterizing implementation challenges.

In chapter 8, Wilson and Clark explore, through two case studies in Montana, efforts to resolve conflict between grizzly bears and humans, a problem that often involves private land adjacent to public, as well as complex partnerships and governance linkages. They discuss technical (e.g., GIS) and other methods as well as participatory processes, illustrating the integration of information and interests. Continuing this theme in chapter 9, Ballard and Fortmann discuss the management of non-timber forest products in the Olympic Peninsula of Washington State, using salal as an illustration. Issues concerning knowledge held outside conventional science, gained here through the authors work with harvesters, are at the core of this case study.

In Chapter 10, Susskind and colleagues examine multi-stakeholder, multidisciplinary environmental science and management problems in detail. They outline the general problem of adversary science, discuss the need for joint fact finding (JFF), and examine consensus building from a JFF perspective. A process and structure for JFF exercises are outlined through a case study on air emissions and cancer risks involving the Northern Oxford County Coalition in Rumford, Maine.

In chapter 11, Wilmsen provides a case study of the Vallecitos forest in New Mexico, drawing on past and present practices of sustained-yield units to focus attention on the social and cultural dimensions of forest management and their intersection with the economic and environmental. This chapter seeks to highlight the challenges and approaches in integrating equity concerns with environmental and economic concerns in community forestry.

Michaels and McCarthy provide a case study of Ontario conservation authorities' information management in chapter 12 (the Water Resource Information Project). They discuss policy development in a watershed management context where stronger measures are not feasible and success depends on collaboration and cooperation. The authors locate the challenges within the broader information and knowledge management literatures, discuss processes, and offer some conclusions and systems contributions.

Finally, in chapter 13, Quinn discusses transboundary natural resource management, including transboundary protected areas, and then outlines lessons learned from practice, drawing on cases from Alberta and Montana's Crown of the Continent system and Europe's Carpathian Ecoregional Planning Initiative.

This diverse and timely collection of perspectives, experiences, and approaches illustrates the complex qualities of contemporary resource and environment management and the integrated nature of natural and human systems. As our knowledge of nature increases, even if only slightly in comparison to what we do not know, we can see, especially now, that there is a greater need than ever before for integrative thinking and comprehensive, consistent integrated approaches to resource and environmental management.

Chapter 1

Integration in Resource and Environmental Management

D. Scott Slocombe and Kevin S. Hanna

INTRODUCTION

Although integration has often been discussed as a component of resource and environmental management (REM), it has rarely been addressed systematically and is hardly ever the primary focus of management. In part this may reflect the complex nature of integration as a component of modern resource and environmental management. While many different approaches and streams of thought support it—from conservation and multiple use to ecosystem approaches, adaptive management, and participatory approaches—there is difficulty in defining such a multi-faceted concept. Certainly integration can mean different things within different broad approaches to REM. In terms of the components of REM, we often think of integrating interests and demands, actors, disciplines, and even the dimensions of sustainability. The applied implications and specific methods of integration can also vary greatly, depending on what is to be integrated or the degree to which integration is desired or appropriate. All of this makes addressing integration in a systematic and comprehensive manner difficult indeed.

The identification of integration as important is not especially new—it was a key part of McHarg's (1969) approach and has resurfaced regularly since, often linked to a systems perspective (e.g., Petak 1980; Barrett 1985; Cairns 1991; Born and Sonzogni 1995; Margerum and Born 1995, 2000). More recently, participation, institutional integration, and communication dimensions have come to the fore. Integration, whatever label it appears under, has emerged as a prominent theme in the policy and management rhetoric of resource and environmental management agencies across North America and abroad. This has been especially true from the perspectives of particular resources and approaches, such as watersheds, forest management, and regional planning (e.g., Campbell and Sayer 2003; Ewert et al. 2004; Hooper et al. 1999; Kennett 2002; Morrison 1995; Thompson and Welsh 1993). Integration is also receiving renewed attention from a scientific research and informational perspective (e.g., Jensen and Bourgeron 2001; Directorate General for Research 2003; van Kerkhoff 2005).

Yet when all is said and done, there can be little doubt that the opposite of integration—fragmentation—remains a substantial obstacle to improving resource and environmental management. Fragmented interests, jurisdictions, ownership, management responsibility, social and ecological systems, information and knowledge all contribute to modern REM challenges. Addressing fragmentation is an essential part of making the ever more necessary fundamental changes to the way that humans manage their relationship with nature and consume natural resources.

In this chapter, we review the development and diversity of integration in resource and environmental management, review challenges to integration, consider the dimensions of integration, and then move on to look at definitions and directions, finishing with some brief conclusions. This is primarily a starting point for reading the chapters that follow, which not only articulate specific challenges in conceptualizing integration, but also illustrate a range of approaches and examples of integration that have emerged in the pursuit of integration in REM.

DEVELOPMENT AND DIVERSITY OF INTEGRATION

There are at least a few different ways to think about the origins and development of integration in resource and environmental management; for example, they can be thought of in terms of the drivers of interest in integration, the main areas of application and development (e.g., forestry and resource management, watershed management, protected areas), or in terms of approaches (e.g., systems, modelling). These are, of course, interrelated, and the following section treats them as such.

Integration has long been a strong theme in natural resources, particularly in forest and water resource management. Since the late nineteenth century, a collection of paradigms has evolved from the early ideas of conservation as wise use—without the quotation marks that phrase currently needs—which included integrated resource management, multiple use, ecosystem approaches, adaptive management, and various decision-making tools, such as environmental impact assessment or policy analysis. The first attempts to achieve integration concentrated on unifying agencies with divergent mandates and cultures or resolving conflicts between preservation and utilization. But as natural resources and the environment have been diminished and the use of nature has become ever more contentious, both resource and environmental management have become more complex, if not more successful. Integrated approaches have also advanced from the early desire to manage public resources to the goal of yielding a range of economic and social welfare benefits and distributing them more equitably.

In their encyclopaedic outline of forest resource management, Duerr et al. (1982) suggested that integrated resource management (IRM) and multiple use have come to mean the same thing. It may be that the most precise differences between the two terms are temporal and purposive; multiple use emphasizes uses or products, while IRM implies a process emphasis. As a distinctive approach to resource management, multiple use became popular in the late 1930s. At the time it was a paradigm primarily centred on considering multiple (economic) benefits in resource management decision making (e.g., Bowes and Krutilla 1989). The integration of agencies and systems, cooperative decision making, and non-economic benefits was not an integral element of all early illustrations of multiple use (e.g., Zivnuska 1961; Davis 1969), even though consideration of such would come to be seen as requisite to realizing multiple benefits.

Integrated resource management became a popular term in the 1960s when the Society of American Foresters sponsored curriculum development projects in which IRM became a topic of discussion. The concept became a component of new approaches to forest resource management (Behan 1990) and spread to other resource settings. But *multiple use* and *IRM* overlap as terms, and in some respects the elements that might differentiate them are not always as clear as those they have in common. Both concepts evolved out of conservation experience into formal frameworks, and like Gifford Pinchot's early articulation of conservation, they initially emphasized the utility of resources and the use of the environment to achieve broader economic and social goals. Social welfare could be supported by maximizing the efficient use of natural resources and more equitably distributing benefits and access. Efficiency was to be served in part by managing resources, or a land base, for a range of products. Natural systems figured into the planning process only in so far as they supported the sustained yield of a benefit. While most would argue that utility and resource production still remain dominant management principles, there have been trends in IRM that emphasize environmental goals, at least from the 1960s on and especially on public lands (e.g., Caldwell 1970; McHarg 1969; Loomis 2002).

The integrated resource and environmental management (IREM) literature has many sources and applications, but includes various land-use planning experiences, notably in British Columbia and Alberta in Canada (e.g., Gunton et al. 2003; Hafso 1992; Kennett 2002; Morrison 1995); large-scale government resource management policy reforms (e.g., Anker 2002; Grinlinton 1992); rural development and planning contexts, often Australian (e.g., Bellamy and Johnson 2000; Hobbs et al. 1993; Hooper et al. 1999; Mitchell and Pigram 1989); and particular resource sectors such as forestry (Thompson and Welsh 1993; Sharitz et al. 1992); coasts and oceans (a huge literature, but, e.g., Burbridge and Humphrey 2003; Cicin-Sain and Knecht 1998; Fast et al. 2005; Kenchington and Crawford 1993; Pernetta and Elder 1993; Power et al. 2000; Van der Weide 1993); and agriculture (e.g., Anderson and Baum 1987). Simultaneously, some emphasize resource survey, scientific, quantitative, and modelling approaches (e.g., Armitage 1995; Liu and Taylor 2002; Wright 1987; Yin and Pierce 1993), and others emphasize process, participation, and institutional approaches (e.g., Anderson and Baum 1987; Ewert et al. 2004; Garcia 1989; Hooper et al. 1999; Stonehouse et al. 1997).

There is also a long history, going back to at least the 1930s and 1940s in parts of Canada and the United States, of integration in river-basin or watershed management, catalysed by the ability of the river basin itself to foster recognition of the interconnected environmental and socio-economic nature of problems and, as in forestry, of the inevitably multi-use and multi-purpose nature of water resource management (cf. White 1969; Schramm 1980; Falkenmark 1981; Sabatier et al. 2005). The literature is very large, but briefly, some approaches emphasize the biophysical issues and technical approaches (e.g., Beck 1997; Cairns and Pratt 1990; Costanza and Jorgenson 2002; Downs et al. 1991), while others emphasize institutional and participatory approaches (e.g., Gardner et al. 1994; Mitchell and Hollick 1993; Lubell et al. 2005). Some authors highlight the tensions between the two (e.g., Hilden 2000) and seek to integrate and balance these two dimensions (e.g., Burton 1995; Calder 1999; Child and Armour 1995; Hooper 2005; Jønch-Clausen and Fugl 2001; Zyl 1995).

Conceptually, too, as in IRM and watershed management, there are approaches and rationales derived from the juxtaposition of development or technological approaches and broader environmental or ecological approaches. Such approaches have tended to emphasize various systems and systematic rationales, often drawing on mathematical models and other quantitative evaluation approaches (e.g., Petak 1980, 1981; Barrett 1985). Here, too, later approaches tend to conjoin this with more attention to planning processes, communication and interaction, and institutional contexts and constraints (e.g., Cairns 1991; Margerum and Born 1995; Born and Sonzogni 1995). This trend has only strengthened over time, with policy, coordination, and participation elements developing steadily (Cairns et al. 1994; Margerum 1999; Margerum and Born 2000; Margerum and Hooper 2001).

Recent thinking with respect to integration has increasingly focused, in varied combinations and emphases, on these two key dimensions: on the one hand, the natural environment and specifically natural systems, and on the other, forms of consultation, participation, and collaboration. Both of these directions have been recently supported and strengthened by resurgent interest in ecosystem approaches and management (e.g., Caldwell 1988; Cortner and Moote 1994; Grumbine 1990; Slocombe 1993, 2001; Kay and Schneider 1994). In rhetoric at least, both resource and environmental management have generally moved beyond the realization that 'simple administrative dictates' are rarely adequate to protect the sustainability of (renewable) resources or the ecosystems that provide them. And with time there has been a gradual, albeit hesitant, initiation of efforts to develop ecologically based policies (Kimmins 1987). Mitchell (1991, 270) commented:

> Given the interrelatedness and complexity, the strategic concern becomes how to deal with the linkages involved within and between various systems (biophysical, human, economic). The response increasingly has been to use an ecosystem approach in which attention turns to consideration of more broadly defined systems rather than focussing upon a specific resource sector such as water, forestry, or minerals. Other terms to describe this approach are holistic, comprehensive, and integrated.

With the new prominence of ecosystem concepts, terms with specific ingredients emerged—not least the idea that the use of natural resources and maintenance of natural systems requires biophysical knowledge. It is acknowledged that environmental degradation is unavoidably tied to the failure of an ecological system to support biological processes at a desired level (Karr 1992). In this vein Norton (1992) proposed that a new paradigm in environmental management is emerging based on the hierarchical complexity of ecosystems and the adoption of the self-organizing character of ecosystems as the centrepiece of environmental management. This idea suggests that fairness to future generations necessitates the protection of the integrity of the large self-organizing systems that provide the context for human life. In such a setting the integrity of ecosystems would become the basis for decision making. Policy would be dictated by the constraints necessary to protect the self-organizing and self-regulating components of the system that together provide the context for decision making (Norton 1992, 2005). But, in practice, integration is still often lacking and such frameworks seem to have assumed the character of a two-stage management system: resource management, which is concerned with economic criteria and the annual cycles of production and consumption; and environmental management, which attempts to apply criteria based on an ecological understanding of a system.

'Good management' increasingly means working with an understanding of ecosystem processes, such as energy cascading (flows), material cycling, and information organizing of ecosystems on the one hand, and of much more complex, inclusive, and collaborative processes of public participation on the other. But current approaches encounter numerous obstacles (discussed further in a later section) and critiques. As Regier (1993) commented, information gathering and analytical approaches centred on ecosystem dynamics and functions have limited utility in practice. Yet information about the operations and requirements of ecosystems and their constituents (what they are composed of) has been important in the improvement of management, even if only by telling us that there are limits to what can be extracted.

Despite the evolution of thought owing to improved ecological knowledge, some resource sectors, especially forestry and water, have doggedly pursued integration largely by trying to achieve some nirvana state of multiple use. But even as integration that is rhetorically based on an ecosystem concept has been advanced, economic benefits are usually the focus and ecology is rarely the primary organizing concept (cf. Bird 1990). Others critique the vagueness and multiple meanings of integration or ecosystem approach concepts (e.g., Bunnell 1995; Scrase and Sheate 2002), the costs and challenges of government reorganizations to foster integrated approaches (e.g., Grigg 1983), the dangers of any approach that suggests we can manage or control natural systems (e.g., Holling and Meffe 1996), the poor implementation of many past IRM efforts, including inadequate funding, goals, and participation (e.g., Thompson 1987), and the degree of importance accorded local history and context in determining the suitability and success of IRM ventures (Walther 1987). It is important to recognize the complex power relationships in integrated management ventures and the often incremental change that results (e.g., Beyers 2001) and to monitor and evaluate such efforts in appropriate ways (see Gottret and White 2001).

Regardless of how integrated resource management is specifically defined, the implication of IRM is that resources can, and should, provide multiple benefits for nature and humans alike. The challenge may lie not so much in defining the concept or accepting its necessity, but rather in putting into operation realistic, consistent, and effective approaches. Management of natural resources, regardless of which type of management framework is used, requires attention to the values held by society. Therefore, identifying values must be an integral part of an IRM framework, regardless of which type of evaluation method is chosen. A consultation process identifies the expectations of resource users and the desired consequences of managing resources. Even within the forms of IRM that explicitly identify ecosystems as a primary consideration, the importance of societal and economic demands cannot be forgotten. As the other chapters in this volume illustrate, integrated approaches must be at least multidimensional and multi-sectoral, reflecting their diverse past and diverse present.

CHALLENGES

With little doubt, the challenges of implementing integrated resource and environmental management are considerable. There are fundamental conceptual and knowledge challenges, numerous practical challenges, and outright barriers. This section briefly reviews some of these. Most fundamentally, integration is affected by the interdependency, complexity, and uncertainty (biophysical, socio-economic, and political) that commonly surround resource and environmental management and are inherent in the ecosystems and societies that produce and use resources (Dorcey 1986; Mitchell 1991; Holling and Meffe 1996). At the level of knowledge and information, there is long-standing recognition of the challenges posed by the disciplinary forms of knowledge, education, and research and by the political, power-related, and bureaucratic impediments to crossing traditional disciplinary boundaries in universities and elsewhere (e.g. Barrett 1985; Cairns and Pratt 1990; Petak 1980). These contribute strongly to problems involving information gaps and availability.

And then there are the challenges of defining the concept, getting beyond theory to clear concepts and criteria for application. Thus, Pearse (1990, 93) commented that while integration ideals such as 'multiple use' may be popular and are frequently extolled as a means of reconciling the growing and often conflicting demands on natural resources, they are also vague concepts that resource managers may have difficulty applying. When, and to what extent, are such approaches *technically feasible* in the accommodation of multiple uses, needs, and demands? In those cases where integrated approaches are technically possible, when is it *desirable* to apply such a model? And based on economic and social grounds, how do we make choices about which resources or uses should be sacrificed for others?

Despite the acceptance of the principle of IRM, progress in implementation of IRM strategies has been largely hesitant and unsystematic. In part this has been because of real obstacles to implementation, often dominated by processes where participants, in a reactive mode, learn as they go, without clear models to follow. The result has been processes where participants tend to be cautious and follow incremental strategies (Mitchell 1990). Somewhat in contrast, there may be a tendency to try to integrate everything or solve too many problems over too large an area. Thus, Mitchell (1991) suggested that many approaches to resource management have been unrealistically broad in scope, whereas a well-defined approach to integration maintains an awareness of the linkages and interrelationships but keeps the process within manageable boundaries—both physically and in terms of issues (i.e., strategic in implementation). Ewert et al. (2004) also highlighted this strategic approach to holistic integration. Equally, however, too small a spatial scale, or too exclusively local a perspective, impedes integration, as decision makers can be trapped by local conditions (e.g., economic and social, such as local or regional employment) into making what turn out to be bad decisions from a long-term perspective (Costanza et al. 1993).

One of the oldest challenges for IRM is fragmentation at the agency level, reflected in poorly integrated institutional arrangements. For any particular resource or environmental system or sector, myriad agencies are responsible for managing it, often working at cross-purposes, encountering resistance from still other agencies and levels of government, and even facing conflicting mandates within one agency. Many other practical problems ultimately derive from this one: data access, monitoring conflicts, boundary issues, jurisdictional battles, funding and personnel shortages, conflicting or duplicate legal requirements, contesting social and economic pressures, weak and/or single-use–oriented legal frameworks, political agendas, and short-term perspectives.

The problems that affect resource and environmental management cannot be characterized as strictly institutional, nor can they be seen solely in terms of bad science or inadequate data. And they are not the exclusive result of social, economic, political, or legal weaknesses. Resource management is affected by the confluence of these realms. And this underscores the multiple tasks, dimensions, and goals of integration that must at least be recognized, if not addressed.

DIMENSIONS OF INTEGRATION

Clearly, one of the issues in seeking integration in resource and environmental management is the need to address the question of what is to be integrated. And equally clearly, there are many possible answers to that question. For the sake of discussion we have identified seven key, interrelated dimensions or areas for integration: disciplines; information; spatial/ecological units; governments; agencies; interests/sectors; and perceptions, attitudes, and values (PAV). Undoubtedly there could be others, and as we will discuss below, some of these could be further broken down. Which of the seven is most important will also vary from case to case—trying to address them all at the same time in the same project is unlikely to be feasible! We will discuss each of them briefly.

Disciplines

Developing understanding of the resources and environment to be managed is a clear first step and has long been recognized as such in planning and related literatures. Developing understanding begs the questions of what knowledge or information is relevant and how it should be organized (that is, what disciplines are relevant—economics, ecology, hydrology, sociology, etc.). The need to bridge and combine disciplines in various ways underlies much of IREM and even broader environmental studies and management literatures. The challenges of such integration are both legion and legendary, and there is a long and voluminous literature on it (e.g., Jantsch 1971; Klein 2001, 2002; Somerville and Rapport 2003; Weingart and Stehr 2000). Integrating disciplines is closely related to integrating information, but here we distinguish the two to emphasize broader structural issues with the former and more detailed process issues with the latter. Illustrating the broader structural perspective, for example, van Kerkhoff (2005) identified 12 themes for integration in research: disciplines, research issues, research and teaching, data, research methods, research organizations, world views, research and application activities, sectors, management/governance arrangements by scale, management/governance organizations by issue focus, and resources. And the European Commission (Directorate General for Research 2003) used a research integration framework distinguishing vertical, horizontal, activity, sectoral, and financial integration.

Information

Many highlight the role of information in REM, in consensus building and changing perceptions and practice on both sides (Hanna 2000), in integrating kinds of knowledge (Holling 1989/90), in educating for both knowledge and the skills needed for its application (Schoenfeld 1981), and even in the long-standing arguments for environmental studies or management's need for 'specialized generalists' or 'generalized specialists' (Francis 1976), although some of these overlap the realm of disciplinary integration as noted. Information, here, can be defined as the specific details of knowledge, from many disciplines, needed for conducting resource and environmental management. And even when we have relevant knowledge about problems, the necessary information is not always readily available or usable. Information comes in many forms, reflecting different sources (e.g., scientific, traditional knowledge), and in different formats (e.g., paper, digital), and seeks to describe different things. Slocombe (2001) reviewed these topics and suggested a range of conceptual, technical, and process means for fostering integration of different forms of information.

Spatial/Ecological Units

The need for more coherent, ecologically sensible management units is an old theme in resource and environmental management (cf. Barrett 1985; Slocombe 1993) and of course a core part of what makes watershed management so sensible. Defining such biophysically and even socio-economically meaningful management units, whether they are watersheds, bioregions, greater ecosystems, ecozones, or something else entirely, tends to require integration across existing, arbitrarily defined administrative units. While approaches to defining these units have become increasingly standard and recognized, not least due to the growth of GIS (geographic information system) and community mapping approaches (cf. Aberley 1993), from an IREM perspective, implementing new multidisciplinary management units raises the particular challenges of integrating multiple governments, at the same and different levels.

Governments

There are numerous ways in which governments may be integrated: several local or regional governments may be brought together within a new, coherent management unit; higher-tier governments may be brought together across major provincial or national borders; or multiple levels of government may simply be brought together within one management unit. All these kinds of government integration tend to face challenges related to different general government goals and objectives, different forms and levels of interest in IREM, lack of resources for integration, and concerns that integration might create more government. A central issue in government integration involves efforts to make management more bottom-up, or community-based, and the desirability of research and other functions being devolved to lower levels of government (for participatory and local knowledge reasons rather than fiscal and ideological downloading reasons). So, for example, there are arguments that participatory research and governance approaches foster improved natural resource management and development better suited to local needs (e.g., Pound et al. 2003).

Agencies

Unlike the broader structural issues pertaining to integrating governments, the challenge of integrating specialized government agencies is one of integrating organizations that have wide geographic mandates but are often disciplinarily and sectorally specialized. This a common theme in the resource and environmental management literature (e.g., Born and Sonzogni 1995; Grinlinton 1992; Margerum 1999; Petak 1981) and remains one of the strongest and hardest to address challenges for integration. Agencies, unlike governments, are not elected and thus cannot be reached directly as easily. Many are also old and entrenched, and changing them takes time and money. Agencies usually play key roles in determining and implementing the laws and policies that provide the broad legal and administrative context for resource and environmental management. While large bureaucracies sometimes accomplish excellent things through the work of particular individuals and projects in specific places, the accomplishment of larger, more systematic changes is very difficult and often depends on changes in the interests and sectors they are linked to.

Interests/Sectors

By interests and sectors, we refer to the often, but not always, economically defined use interests in resources and environment—for example, forestry, mining, manufacturing, transportation, recreation, parks, and conservation. Integrating different interests depends tremendously on their mutual compatibility, the inherent environmental and other effects, the legal and policy framework, and other incentives for compromise and accommodation. Integration here is also closely associated with issues involving public and private interests and rights and the need to balance (or integrate) them. While sectors, like agencies, may sometimes be well integrated and balanced in particular projects, places, and times, broader or more universal integration remains problematic, perhaps not reasonably achievable, and highly dependent on broader political, economic, and institutional factors, which can be deeply influenced by individual and societal perceptions, attitudes, and values.

Perceptions, Attitudes, and Values (PAV)

At the most fundamental level, in particular places and times, it may be argued that what must be integrated are the different perceptions, attitudes, and values held by different individuals, organizations, and institutions. Different perceptions, attitudes, and values can affect understandings and interpretations of information, goals and objectives, and history, as well as what forms of integration of management units, governments, agencies, and interests may be feasible in the future. Recognizing and seeking to work with PAV is a big part of the rationale for more participatory management at local levels and more formal conflict resolution and negotiation approaches at higher levels. Yet PAV remains one of the most intractable and contentious areas of integration, relatively little written about, at least directly. Integration must deal carefully, sensitively, and inclusively with the range of perceptions, attitudes, and values to be found in regions that are the focus of IREM initiatives (cf. Reed 2003).

DEFINITIONS AND DIRECTIONS

Creating a universal definition for something as complex and multi-faceted as integrated resource and environmental management is difficult if not impossible—not the least of the reasons we prefer to speak of integration in REM rather than of IREM (while admittedly resorting to the latter as a convenient shorthand). However, there is an argument that without a decent definition, it becomes difficult to establish goals, design implementation strategies, and monitor progress (Mitchell 1990). In addition, a good definition can help us develop shared understanding of how to define and implement an integrated approach and how to adapt, or reform, decision-making processes to realize it. A good definition can help to explain what is meant by an integrated approach, how we can know one when we see one, and how we determine that such an approach is appropriate to the context and problem. A collection of principles is apparent (discussed below), but even these can vary between the devotees of a specific resource or between the devotees of specific tools—such as impact assessment, ecosystem-based management, or even benefit-cost analysis.

In practice, integration has evolved from approaches applied more commonly to resources such as timber or water to those that embrace broader concepts that are transferable to a broad range of environmental problems and settings. This reflects the gradual growth in understanding of the social, cultural, and economic interrelatedness of environmental use (Gibson 1993). In an ideal sense, integrated processes are broadly interactive, integrating the multiple demands and values of all resource users into management strategies. But it is difficult to say that there is a universal model of integration. Thus, in the following sections we review some of the definitions of integration, many of which reflect distinctions and themes noted in the development and evolution section above.

Mitchell (1986, 13) provided a definition of IRM that accents some common procedural principles: 'Although the choice of specific descriptors will vary, the usual idea associated with integrated resource management is the sharing and coordination of the values and inputs of a broad range of agencies, publics and other interests when conceiving, designing, and implementing policies, programs or projects.'

Similarly, Walther (1987, 439–40) defined integration and integrated resource management in terms of increasing cooperation and communication among decision makers and developing a structured and consistent forum for policy development and implementation: '[I]ntegration is a process of increasing organization and order in a system. Most IRM projects approach integration by improving communication and applying the concept of cooperative decision-making among experts of sectoral interest groups.' Walther's definition is compelling, because it offers a way of measuring the existence or success of IRM. Though such definitions suggest a multi-stakeholder and consensual approach to decision making, integration has also been defined in very specific, utilitarian terms that stress the role of an individual, or an individual agency, as the sole decision maker. While we can say that approaches to defining integration certainly vary, as Child and Armour (1995, 116–19) wrote, there are commonalities. The following are one take on these: multiple means and multiple purposes (Mitchell 1990; White 1969), multi-sectoral blending, the incorporation of multiple professions and perspectives into planning, public participation, and the achievement of accommodation and compromise. Integration implies cooperative decision making and cooperative planning.

In a different take on resource management, Lang (1986) suggested that conventional planning is dominated by three approaches: comprehensive rationality, incrementalism, and a combination of these two. Alternatively, Lang suggested that planning processes develop strategic, interactive, and multiple perspectives on resource management and environmental problems in order to achieve results that not only reflect an integrated approach, but also move towards addressing the needs and values of different stakeholders. Lang's discussion highlighted the continuing need for strategic and consistent approaches to managing natural resources, on an integrated basis, to reduce conflict and decrease uncertainty. These are oft-reiterated themes, with a few others added, such as holistic, comprehensive, inclusive, and interconnected themes (cf. Mitchell 1986; Born and Sonzogni 1995; Margerum 1997; Ewert et al. 2004).

While models and definitions of integration vary, two themes have often dominated—comprehensiveness and strategy. Comprehensiveness emphasizes broad consultation, participation among all stakeholders, and the consideration of an extensive range of issues and solutions. But despite its integrative elements, comprehensiveness poses feasibility problems and may consume excessive time and resources. A strategic or tractable approach is based on a more defined process of inter-agency cooperation with more limited jurisdictions and physical boundaries (Born and Sonzogni 1995; Hanna 2000). The scope of comprehensiveness may often be simply too large. Strategic approaches concentrate on developing a structured and consistent forum for policy development and implementation. They focus on a more defined range of issues and options and tend to emphasize practical planning solutions while concentrating on key components and linkages within a problem area (Mitchell 2002). Strategic approaches may also be less inclusive, a by-product of the focus imperative. Under either defining label, integration is a process of increasing organization and order in a decision-making system and across interests (Walther 1987). Newer approaches emphasize the complexity of the planning environment and seek to foster adaptive learning approaches that are more participatory, embrace key principles such as multi-scale analysis, emphasize governance rather than government, and draw on tools such as systems modelling, policy analysis, decision and negotiation support systems, and impact assessment (cf. Brunner et al. 2005; Ewert et al. 2004; Sayer and Campbell 2001).

Integration in both the strategic and comprehensive contexts assumes the image of deliberative process where negotiation assumes an integral role. Indeed, integration may be very much about the creation of a setting within which stakeholders create a social basis for action. The image of deliberative process is particularly appropriate to the intent of integration. The broad concept of integration, irrespective of dominant themes or contexts, is embodied in a language of coordination. Within this ambit, integration can be described as a process based on reaching a mutual understanding of issues and approaches, the coordination of action, and the socialization of participants (agencies, the public, and non-governmental organizations). Socialization is ultimately expressed in the language of REM through plans, policy statements, and the common definitions of issues, problems, and solutions.

The integrated setting may become a new form of institutional reality. From a critical perspective, integration may create a new set of rules for interaction where power is no longer exercised individually by agencies, but within a new integrative setting—though agencies are not always explicitly aware of the impact of the new rules of deliberation. Negotiation imparts influence and power to the process, and participants use it to seek mutual understanding and shared information rather than advantage through conflict. This suggests that successful integration leads to relative *homogeneity* among participants in terms of how they define the problem, the solutions, program and policy impacts, or the state of the environment, and even the extent to which they may adopt a common program language. In some contexts, increasing homogeneity may be an issue in and of itself.

There may be no single model, but there is an implicit consensus that integration means the reduction of agency fragmentation, the injection of cooperation into organizational culture, the use of diverse information sources and knowledge, participation, and accommodation and consensus rather than conflict (Child and Armour 1995; Lang 1986). Integration implies cooperative decision making and cooperative planning—it is a process. But all this can be difficult to achieve. Many institutional settings reinforce self-interest rather than support integration and cooperation. There can also be the perception that power may be lost through integrated planning, and indeed it may be. From a critical theoretical perspective, these elements highlight the potential for implementation challenges, since even modest forms of integration require a change in power relationships, which some might see as being a primary obstacle in any implementation process.

In putting together this book, we first needed a working definition of integration, one that would reflect the complexity that imbues not only contemporary resource and environmental management, but also echoes the evolution in thinking that now surrounds integration. As a starting point we sent the authors an earlier version of this malleable definition:

In practice, integration in resource and environmental management has evolved from the tools applied commonly to managing natural resources such as timber or water, to conceptually broader and more sophisticated approaches—ones that are increasingly transferable to a broad range of environmental problems and settings. This evolution reflects the gradual growth in understanding of the interrelatedness of the social, cultural, economic, and ecological dimensions of resource and environmental problems. In an ideal sense, integrated resource and environmental management draws on scientific and other forms of knowledge, information and other forms of technology, and collaborative and other processes to foster better resource and environmental management through improved integration of some or all of, but not limited to, the following dimensions: disciplines, information, spatial/ecological units, governments, agencies, interests/sectors, and perceptions, attitudes, and values.

Although this definition is broad, the activity of defining and implementing integration has come to reflect an increasingly intricate view of how REM must be implemented; how integration can be defined, applied, and measured; and how the environmental, social, and political challenges inherent in realizing integrated approaches to resource and environmental management can be addressed.

CONCLUSION

This chapter has sought to give a sense of the origins, development, themes, and issues in the effort to foster integration in resource and environmental management. It takes a book to do this fully, so the preceding can be but a taste of the history and complexity. And this introduction has also been a fairly high-level overview. The specifics of implementation, of what works, are also very diverse and have been addressed by many writers, including ourselves. Table 1.1 presents a sampling of common suggestions for making integrated resource and environmental management work (after, among others, Margerum 1999; Born and Sonzogni 1995; Hooper et al. 1999). The chapters that follow will provide many concrete illustrations of these and other ideas, practices, and prerequisites.

There are other topics that could be explored, newer approaches and extensions of current ones, such as moving beyond systems to complex systems ideas (e.g., Geldof 1995; Grzybowski and Slocombe 1988; Slocombe 2001; Yevjevich 1995) or exploring more deeply the institutional (cf. Connor and Dovers 2004), or even societal, changes that are needed for lasting integration to occur. But those will have to be left for another venue. We hope that the ideas here and in the chapters that follow will help others pursue integration in particular places, times, and projects.

This introductory chapter foreshadows and contextualizes the chapters to come. Those chapters further illustrate the multiple meanings of integration in resource and environmental management, and reflect the complex, varied, and continuously evolving process of characterizing resource and environmental management problems and the solutions that can be employed to improve the state of management.

TABLE 1.1 Some commonly identified operational factors and elements for successful integrated resource and environmental management

Systems Approach	Incentives for Reaching a Solution
Context	Clear decision rules
Supportive political climate and politicians	Intervention options
Supportive legal and policy framework	Equitable access to necessary information
Staff and financial resources	Simulation modelling
Time	Information and decision-support systems
Local- and high-level leadership	Strengthening regional and local capacity
Interdisciplinary teams	Encouraging best-management practices
Open organizations	Communication, learning, trust-building opportunities
Coordination mechanisms and structures	Information and education programs
Willing major stakeholder participation	Monitoring and evaluation programs

REFERENCES

Aberley, D., ed. 1993. *Boundaries of home: Mapping for local empowerment.* Gabriola Island, BC: New Society Publishers.

Anderson, E.W., and R.C. Baum. 1987. Coordinated resource management planning: Does it work? *Journal of Soil and Water Conservation* 42 (3): 161–6.

Anker, H. 2002. Integrated resource management: Lessons for Europe? *European Environmental Law Review* 11 (7): 199–209.

Armitage, D. 1995. An integrative methodological framework for sustainable environmental planning and management. *Environmental Management* 19 (4): 469–79.

Barrett, G.W. 1985. A problem-solving approach to resource management. *BioScience* 35 (7): 423–7.

Beck, M.B. 1997. Applying systems analysis in managing the water environment: Towards a new agenda. *Water Science and Technology* 36 (5): 1–17.

Behan, R.W. 1990. Multiresource forest management: A paradigmatic challenge to professional forestry. *Journal of Forestry* 88:12–18.

Bellamy, J., and A. Johnson. 2000. Integrated resource management: Moving from rhetoric to practice in Australian agriculture. *Environmental Management* 25 (3): 265–80.

Beyers, J.M. 2001. Model forests as process reform: Alternative dispute resolution and multistakeholder planning. In M. Howlett (ed.), *Canadian forest policy: Adapting to change*, 172–202. Toronto: University of Toronto Press.

Bird, I.D. 1990. The potential for integrated resource management with intensive or extensive forest management: Reconciling vision with reality. *Forestry Chronicle*, October, 444–6.

Born, S.M., and W.C. Sonzogni. 1995. Integrated environmental management: Strengthening the conceptualization. *Environmental Management* 19 (2): 167–81.

Bowes, M.D., and J.V. Krutilla. 1989. *Multiple-use management: The economics of public forestlands*. Washington, DC: Resources for the Future.

Brunner, R.D., T.A. Steelman, L. Coe-Juell, C.M. Cromley, C.M. Edwards, and D.W. Tucker. 2005. *Adaptive governance: Integrating science, policy, and decision making*. New York: Columbia University Press.

Bunnell, F. 1995. Toto, This isn't Kansas: Changes in integrated forest management. In *Selected papers from the Integrated Forest Management: Proceedings of Model Forest Network Workshop, 3–5 October 1994, Sussex, New Brunswick*, 1–14. Ottawa, ON: Canadian Forest Service.

Burbridge, P., and S. Humphrey. 2003. Introduction to special issue on the European Demonstration Programme on Integrated Coastal Zone Management. *Coastal Management* 31 (2): 121–6.

Burton, J. 1995. A framework for integrated river basin management. *Water Science and Technology* 32 (5/6): 139–44.

Cairns, John, Jr. 1991. The need for integrated environmental management. In J. Cairns Jr. and T.V. Crawford (eds), *Integrated environmental management*, 5–20. Chelsea, MI: Lewis Publishers.

Cairns, J., Jr., and J.R. Pratt. 1990. Integrating aquatic ecosystem resource management. In R. McNeil and J.E. Windsor (eds), *Innovations in river basin management*, 265–80. Cambridge, ON: Canadian Water Resources Association.

Cairns, J., Jr., T.V. Crawford, and H. Salwasser (eds), 1994. *Implementing integrated environmental management*. Blacksburg, VI: Virginia Polytechnic Institute and State University, University Center for Environmental and Hazardous Materials Studies.

Calder, Ian R. 1999. *The blue revolution: Land use and integrated water resources management*. London: Earthscan Publications.

Caldwell, L.K. 1970. The ecosystem as a criterion for public land policy. *Natural Resources Journal* 10 (2): 203–21.

——— (ed.). 1988. Implementing an ecological systems approach to basinwide management. In *Perspectives on ecosystem management for the Great Lakes: A reader*, 1–29. Albany, NY: SUNY Press.

Campbell, B., and J. Sayer (eds), 2003. *Integrated natural resource management: Linking productivity, the environment and development*. Wallingford: CABI Publishing.

Centre for Educational Research and Innovation (CERI). 1972. *Interdisciplinarity: Problems of teaching and research in universities*. Paris: OECD.

Child, M., and A. Armour. 1995. Integrated water resource planning in Canada: Theoretical considerations and observations from practice. *Canadian Water Resources Journal* 20 (2): 115–26.

Cicin-Sain, B., and R.W. Knecht. 1998. *Integrated coastal and ocean management: Concepts and practices*. Washington, DC: Island Press.

Connor, R., and S. Dovers. 2004. *Institutional change for sustainable development*. Cheltenham, UK: Edward Elgar Publishing.

Cortner, H.J., and M.A. Moote. 1994. Trends and issues in land and water resources management: Setting the agenda for change. *Environmental Management* 18 (2): 167–73.

Costanza, R., and S.E. Jorgensen (eds). 2002. *Understanding and solving environmental problems in the 21st century: Towards a new, integrated hard problem science.* Oxford, UK: Elsevier Science.

Costanza, R., W.M. Kemp, and W.R. Boynton. 1993. Predictability, scale, and biodiversity in coastal and estuarine ecosystems: Implications for management. *Ambio* 22 (2–3): 88–96.

Darling, F.F., and R.F. Dasmann. 1969. The ecosystem view of human society. *Impact of Science on Society* 19 (2): 109–21.

Davis, K.P. 1969. What multiple forest land use and for whom? *Journal of Forestry* 67:719–21.

Directorate General for Research. 2003. Provisions for implementing integrated projects: Background document. European Commission.

Dorcey, A.H.J. 1986. *Bargaining and the governance of Pacific coastal resources: Research and reform.* Vancouver: Westwater Research Centre.

Downs, P.W., K.J. Gregory, and A. Brookes. 1991. How integrated is river basin management? *Environmental Management* 15 (3): 299–309.

Duerr, W.A., D.E. Teeguarden, N.B. Christiansen, and S. Guttenberg (eds). 1982. *Forest resource management.* Corvallis, OR: O.S.U. Bookstores.

Ewert, A.W., D.C. Baker, and G.C. Bissix. 2004. *Integrated resource and environmental management: The human dimension.* Wallingford, UK: CABI Publishing.

Falkenmark, M. 1981. Integrated view of land and water. *Geografiska Annaler* 63 A (3–4): 261–71.

Fast, H., D.B. Chiperzak, K.J. Cott, and G.M. Elliott. 2005. Integrated resource management planning in Canada's western arctic: An adaptive consultation process. In F. Berkes et al. (eds), *Breaking ice: Renewable resource and ocean management in the Canadian North*, 95–117. Calgary: University of Calgary Press.

Francis, G.R. 1976. An overview and summing up—The Rungstad Conference. In *Environmental problems and higher education*, 48–78. Paris: OECD-CERI.

Garcia, Margot W. 1989. Forest Service experience with interdisciplinary teams developing integrated resource management plans. *Environmental Management* 13 (5): 583–92.

Gardner, J., K. Thomson, and M. Newson. 1994. Integrated watershed/river catchment planning and management: A comparison of selected Canadian and United Kingdom experiences. *Journal of Environmental Planning and Management* 37 (1): 53–67.

Geldof, G.D. 1995. Adaptive water management: Integrated water management on the edge of chaos. *Water Science and Technology* 32 (1): 7–13.

Gibson, R.B. 1993. Environmental assessment design: Lessons from the Canadian experience. *Environmental Professional* 15:12–4.

Gottret, M.A.V.N., and D. White. 2001. Assessing the impact of integrated natural resource management: Challenges and experiences. *Conservation Ecology* 5 (2): 17. http://www.consecol.org/vol5/iss2/art17.

Grigg, N.S. 1983. Is integrated environmental management feasible? *Journal of Professional Issues in Engineering* 109 (2): 71–80.

Grinlinton, D.P. 1992. Integrated resource management—a model for the future. *Environmental and Planning Law Journal*, 4–19 February.

Grzybowski, A.G.S., and D.S. Slocombe. 1988. Self-organization theories and environmental management: The case of South Moresby. *Environmental Management* 12 (4): 463–78.

Grumbine, E. 1990. Protecting biological diversity through the greater ecosystem concept. *Natural Areas Journal* 10 (3): 114–20.

Gunton, T.I., J.C. Day, and P.W. Williams (eds). 2003. Special issue: Collaborative planning and sustainable resource management: The North American experience. *Environments* 31 (2): 1–72.

Hafso, Toni. 1992. Landscape themes in integrated resource planning—'The Coal Branch Experience.' In G.B. Ingram and M.R. Moss (eds), *Proceedings of the 2nd Canadian Society for Landscape Ecology and Management Meeting*, 201–6. Morin Heights, QC: Polyscience Publications.

Hanna, K.S. 2000. The paradox of participation and the hidden role of information: A case study. *APA Journal* 66 (4): 398–410.

Hilden, M. 2000. The role of integrating concepts in watershed rehabilitation. *Ecosystem Health* 6 (1): 39–50.

Hobbs, R.J., D.A. Saunders, and G.W. Arnold. 1993. Integrated landscape ecology: A Western Australian perspective. *Biological Conservation* 64 (3): 231–8.

Holling, C.S. 1989/90. Integrating science for sustainable development. *Journal of Business Administration* 19 (1/2): 73–83.

Holling, C.S., and G.K. Meffe. 1996. Command and control and the pathology of natural resource management. *Conservation Biology* 10 (2): 328–37.

Hooper, B. 2005. *Integrated river basin governance: Learning from international experience.* London, UK: IWA Publishing.

Hooper, B.P., G.T. McDonald, B. Mitchell. 1999. Facilitating integrated resource and environmental management: Australian and Canadian perspectives. *Journal of Environmental Planning and Management* 42 (5): 747–66.

Jantsch, E. 1971. Inter- and trans-disciplinary university: A systems approach to education and innovation. *Ekistics* 32:133–9.

Jønch-Clausen, T., and J. Fugl. 2001. Firming up the conceptual basis of integrated water resources management. *International Journal of Water Resources Development* 17 (4): 501–10.

Jensen, M.E., and P.S. Bourgeron (eds). 2001. *A Guidebook for integrated ecological assessments.* New York: Springer Verlag.

Karr, J.R. 1992. Ecological integrity: Protecting earth's life support systems. In R. Costanza, B.G. Norton, and B. Hakell (eds), *Ecosystem health: New goals for environmental management.* Washington, DC: Island Press.

Kay, J.J., and E. Schneider. 1994. Embracing complexity: The challenge of the ecosystem approach. *Alternatives* 20 (3): 32–9.

Kenchington, R., and D. Crawford. 1993. On the meaning of integration in coastal zone management. *Ocean and Coastal Management* 21:109–27.

Kennett, S.A. 2002. *Integrated resource management in Alberta: Past, present and benchmarks for the future.* University of Calgary CIRL Occ. Paper 11, Calgary.

Kimmins, J.P. 1987. *Forest ecology.* New York: Macmillan.

Klein, J.T. (ed.). 2001. *Interdisciplinarity: History, theory and practice.* Detroit, MI: Wayne State University Press.

——— (ed.). 2002. *Transdisciplinarity: Joint problem solving among science, technology, and society—An effective way for managing complexity.* Birkhäuser.

Lang, R. (ed.). 1986. *Integrated approaches to resource planning and management.* Calgary: University of Calgary Press.

Liu, J., and W. Taylor (eds). 2002. *Integrating landscape ecology into natural resource management.* Cambridge, UK: Cambridge University Press.

Loomis, J. 2002. *Integrated public lands management: Principles and applications to national forests, parks, wildlife refuges, and BLM lands.* 2nd edn. New York: Columbia University Press.

Lubell, M., P.A. Sabatier, A. Vedlitz, W. Focht, Z. Trachtenberg, and M. Matlock. 2005. Conclusions and recommendations. In *Swimming upstream: Collaborative approaches to watershed management,* 261–96. Cambridge, MA: MIT Press.

McHarg, I.L. 1969. *Design with nature.* New York: Doubleday/Natural History Press.

Margerum, R. 1997. Integrated approaches to environmental planning and management. *Journal of Planning Literature* 11 (4): 459–75.

——— 1999. Integrated environmental management: The foundations for successful practice. *Environmental Management* 24 (2): 151–66.

Margerum, R.D., and S.M. Born. 1995. Integrated environmental management: Moving from theory to practice. *Journal of Environmental Planning and Management* 38 (3): 371–91.

——— 2000. A co-ordination diagnostic for improving integrated environmental management. *Journal of Environmental Planning and Management* 43 (1): 5–21.

Margerum, R.D., and B.P. Hooper. 2001. Integrated environmental management: Improving implementation through leverage point mapping. *Society & Natural Resources* 14 (1): 1–19.

Mitchell, B. 1986. The evolution of integrated resource management. In R. Lang (ed.), *Integrated approaches to resource planning and management,* 13–26. Calgary: University of Calgary Press.

——— 1990. Integrated water management. In *Integrated water management: International perspectives and experiences.* New York: Belhaven.

——— 1991. BEATing conflict and uncertainty in resource management and development. In B. Mitchell (ed.), *Resource management and development,* 268–85. Don Mills, ON: Oxford University Press.

——— 2002. *Resource and environmental management.* 2nd edn. Harlow, UK: Prentice Hall.

Mitchell, B., and M. Hollick. 1993. Integrated catchment management in Western Australia: Transition from concept to implementation. *Environmental Management* 17 (6): 735–43.

Mitchell, B., and J.J. Pigram. 1989. Integrated resource management and the Hunter Valley Conservation Trust, NSW, Australia. *Applied Geography* 9:196–211.

Mitchell, B., and W.R.D. Sewell (eds). 1981. *Canadian resource policies: Problems and prospects.* Toronto: Methuen.

Morrison, N.R. 1995. Integrated resource management planning: A vehicle for environmental assessment. *Plan Canada* 35 (2): 34–6.

Morrison, T.H., G.T. McDonald, and M.B. Lane. 2004. Integrating natural resource management for better environmental outcomes. *Australian Geographer* 35 (3): 243–58.

Norton, B.G. 1992. A new paradigm for environmental management. In R. Costanza, B.G. Norton, and B. Hakell (eds), *Ecosystem health: New goals for environmental management,* 23–41. Washington, DC: Island Press.

——— 2005. *Sustainability: A philosophy of adaptive ecosystem management.* Chicago: University of Chicago Press.

Palmer, C.L. 2001. *Work at the boundaries of science: Information and the interdisciplinary research process.* Dordrecht, Netherlands: Kluwer Academic.

Pearse, P.H. 1990. *Forestry economics.* Vancouver: University of British Columbia Press.

Pernetta, J., and D. Elder. 1993. *Cross-sectoral, integrated coastal area planning (CICAP): Guidelines and principles for coastal area development.* Gland, Switzerland: IUCN Marine and Coastal Areas Programme.

Petak, W.J. 1980. Environmental planning and management: The need for an integrative perspective. *Environmental Management* 4 (4): 287–95.

───── 1981. Environmental management: A system approach. *Environmental Management* 5 (3): 213–24.

Pound, B., S. Snapp, C. McDougall, A. Braun (eds). 2003. *Managing natural resources for sustainable livelihoods: Uniting science and participation.* London: Earthscan; Ottawa: International Development Research Centre.

Power, J., J. McKenna, M.J. MacLeod, A.J.G. Cooper, and G. Convie. 2000. Developing integrated participatory management strategies for Atlantic dune systems in County Donegal, Northwest Ireland. *Ambio* 29 (3): 143–9.

Reed, Maureen G. 2003. *Taking stands: Gender and the sustainability of rural communities.* Vancouver: University of British Columbia Press.

Regier, H.A. 1993. The notion of natural and cultural integrity. In S. Woodley, J. Kay, and G. Francis (eds), *Ecological integrity and the management of ecosystems*, 3–18. St Lucie: St Lucie Press.

Sabatier, P.A., C. Weible, and J. Ficker. 2005. Eras of water management in the United States: Implications for collaborative watershed approaches. In P.A. Sabatier, W. Focht, M. Lubell, Z. Trachtenberg, A. Vedlitz, and M. Matlock (eds), *Swimming upstream: Collaborative approaches to watershed management*, 23–52. Cambridge, MA: MIT Press.

Sayer, J.A., and B.M. Campbell. 2001. Research to integrate productivity enhancement, environmental protection and human development. *Conservation Ecology* 5 (2). Online 32 pp.

Schoenfeld, C. 1981. Educating for integrated resource management. *Environmentalist* 1 (2): 117–22.

Schramm, G. 1980. Integrated river basin planning in a holistic universe. *Natural Resources Journal* 20 (4): 787–806.

Scrase, J.I., and W.R. Sheate. 2002. Integration and integrated approaches to assessment: What do they mean for the environment? *Journal of Environmental Policy and Planning* 4 (4): 275–94.

Sharitz, R.R., L.R. Boring, D.H. Van Lear, and J.E. Pinder, III. 1992. Integrating ecological concepts with natural resource management of southern forests. *Ecological Applications* 2 (3): 226–37.

Slocombe, D.S. 1993. Environmental planning, ecosystem science, and ecosystem approaches for integrating environment and development. *Environmental Management* 17 (3): 289–303.

───── 2001. Integration of biological, physical, and socio-economic information. In M.E. Jensen and P.S. Bourgeron (eds), *A guidebook for integrated ecological assessments*, 119–32. New York: Springer Verlag.

Somerville, M., and D. Rapport (eds). 2003. *Transdisciplinarity: Recreating integrated knowledge.* Montreal and Kingston: McGill-Queen's University Press.

Stonehouse, D.P., C. Giraldex, and W. van Vuuren. 1997. Holistic policy approaches to natural resource management and environmental care. *Journal of Soil and Water Conservation* 52 (1): 22–5.

Thompson, I.D. 1987. The myth of integrated wildlife/forestry management. *Queen's Quarterly* 94 (3): 609–21.

Thompson, I.D., and D.A. Welsh. 1993. Integrated resource management in boreal forest ecosystems: Impediments and solutions. *Forestry Chronicle* 69 (1): 32–9.

Van der Weide, J. 1993. A systems view of integrated coastal management. *Ocean & Coastal Management* 21:129–48.

van Kerkhoff, L. 2005. Integrated research: Concepts of connection in environmental science and policy. *Environmental Science and Policy* 8:452–63.

Walther, P. 1987. Against idealistic beliefs in the problem solving capacities of integrated resource management. *Environmental Management* 11 (4): 439–46.

Weingart, P., and N. Stehr (eds). 2000. *Practising interdisciplinarity.* Toronto: University of Toronto Press.

White, G.F. 1969. *Strategies of American water management.* Ann Arbour: University of Michigan.

Wright, R.L. 1987. Integration in land research for Third World development planning: An applied aspect of landscape ecology. *Landscape Ecology* 1 (2): 107–17.

Yin, Y., and J.T. Pierce. 1993. Integrated resource assessment and sustainable land use. *Environmental Management* 17 (3): 319–27.

Yevjevich, V. 1995. Effects of area and time horizons in comprehensive and integrated water resources management. *Water Science and Technology* 31 (8): 19–25.

Zivnuska, J.A. 1961. The multiple problems of multiple use. *Journal of Forestry* 59:555–60.

Zyl, F.C. van. 1995. Integrated catchment management: Is it wishful thinking or can it succeed? *Water Science and Technology* 32 (5/6): 27–35.

Chapter 2

An Overview of Integration in Resource and Environmental Management

Bruce Mitchell and Dan Shrubsole

INTRODUCTION

The management of water and land resources, an issue involving the consideration of many natural and human factors, has become increasingly complex. As a result, land and water management has often been a testing ground for initiatives to coordinate various levels of government and different user groups. *Integrated resource and environmental management* (IREM) has been, and is increasingly being, advocated to meet this challenge. *Integration* is neither a new concept nor one devised to address new issues (White 1957; Weber 1964; Schramm 1980; Mitchell 1986, 1990), and it offers attractive benefits. However, IREM is often used loosely and with different meanings. For its full benefits to be realized, we need to understand what it does and does not involve. Such an understanding should be based on the people who use resources, the motivation for their use, and the roles that public policy and agencies play in influencing use. It should also consider outcomes and impacts related to IREM.

In this chapter, we examine the evolution of IREM, first by describing key concepts and then by reviewing its interpretation and application both on a global scale and by selected public resource agencies in the United States and Canada. We conclude by focusing on key challenges for IREM and on possible innovations to address these challenges.

KEY CONCEPTS

In the *World Book Dictionary*, *integration* is defined as 'to make into a whole, to put or bring together parts into a whole'. For integration to occur, *collaboration* ('act of working together') and *coordination* ('harmonious adjustment or working together', or 'arrangement in proper order' or relationships) normally are necessary. Furthermore, integration is a means, not an end. Therefore, the use of IREM should be preceded by a shared vision about a desired future condition or state. Without such direction, it is difficult to determine which parts need to be made or brought together into a whole and who should be working together to arrange proper order and relationships.

The rationale for using integration to help achieve a vision is that it allows the desired future condition to be achieved *effectively* (produce the desired effects), *efficiently* (produce the desired effects without waste of time and energy), and *equitably* (ensure that the benefits and costs of the desired effects are distributed fairly among people in space and time). IREM is usually advocated because of its potential to contribute with respect to all three criteria. It contributes to effectiveness by helping to ensure that different needs and opportunities are considered and incorporated into plans and activities, to efficiency by helping to ensure that the actions of one agency or organization do not undo the actions of another agency, and to equity by forcing the consideration of different values and interests of various stakeholders.

THE INTERNATIONAL ROOTS OF IREM

The early history of the global environmental movement was dominated by activities focused initially on regional and larger geographic scales, both requiring collaboration and coordination. One early effort was the North American Conservation Conference, held in Washington, DC, in 1909. Delegates from Canada, Newfoundland, Mexico, and the United States discussed conservation principles. A subsequent world conference was planned, and invitations were issued to 58 countries, but this meeting was cancelled when President Theodore Roosevelt's administration was replaced by William H. Taft's (McCormick 1995). After that, there was little significant collaboration at a global scale until the 1972 United Nations Conference on the Human Environment. This was followed by the UN Conference on Environment and Development (1992) and the World Summit on Sustainable Development (2002), both considered below.

The United Nations Conference on the Human Environment, held in Stockholm, was an important event in the evolution of IREM because for the first time the social, political, and economic dimensions of environmental problems were discussed together at an international level. The conference aimed to 'create a basis for comprehensive consideration within the United Nations of the problems of the human environment' and 'to focus the attention of governments and public opinion in various countries on the importance of the problem' (McCormick 1995, 107)—all of which would require collaboration, coordination, and, ultimately, integration.

The essence of what today would be called IREM was contained in the *Stockholm Declaration on Human Environment*, which presented 26 principles. McCormick (1995, 126) classified the principles into five groups, of which the following three required collaboration and coordination:

- Development and environment concerns should go together, and less developed countries should be given every assistance and incentive to promote rational environmental management.
- Each state should establish its own standards of environmental management and exploit resources as it wished, but should not endanger other states. There should be international cooperation aimed at improving the state of the environment.

- Science, technology, education, and research should all be used to promote environmental protection.

Implicit in the principles is the idea that humans are part of the environment—not apart from it—and that nations' economic development is closely tied to natural resource management. The principles respect nations' ability to develop resources, although such development must be tempered by consideration for impacts on other nations. They also promote the use of science, technology, education, and research. The creation of the United Nations Environment Programme (UNEP) was one of the most visible outcomes of the meeting, and it continues to promote awareness of and effective action on environmental issues. However, this orientation focuses attention on the natural resources themselves, rather than on human perception and use of resources. Following from this perspective, IREM emphasizes that it is important to understand the biophysical aspects of resource systems in order that they might support multiple uses, particularly if science and technology improve the efficiency of resource use.

Although several international agreements, such as the Law of the Sea, were signed after 1972, it became increasingly apparent that many environmental problems were intricately related to economic and social issues such as overpopulation and human consumption of resources. Partially in response to this awareness, the UN established the World Commission on Environment and Development (Brundtland Commission) 'to re-examine the critical issues of environment and development and to formulate innovative, concrete and realistic proposals to deal with them' (Grubb et al. 1993, 7). The concept of *sustainable development* was proposed to meet this challenge, and it came to be understood to mean development that 'meets the needs of the present without compromising the ability of future generations to meet their own needs' (WCED 1987, 43). A fundamental belief of sustainable development is that economic and ecological concerns are interdependent and need to be integrated. In the words of the Brundtland Commission, 'economy is not just about the production of wealth, and ecology is not just about the protection of nature; they are both equally relevant for improving the lot of humankind' (WCED 1987, 38). People expect the environment to sustain life, support community and economic activities, and look beautiful, but balancing these expectations while meeting people's needs is a challenge, one that lies at the core of IREM. The commission (WCED 1987, 62) maintained that if this balance is to be realized, institutional arrangements must be reformed. It also advocated the need for environmental policies to be given the same level of importance as finance, energy, food, and other government policy areas.

The UN Conference on the Environment and Development (Earth Summit or Rio Summit) was convened, in part, to take stock of how nations were implementing sustainable development. In June 1992, 178 countries and over 500 different groups attended the summit that would produce five key agreements: the Framework Convention on Climate Change, the Convention on Biodiversity, Agenda 21, the Rio Declaration on Environment and Development, and the Forest Principles. The latter three are relevant for this chapter. Agenda 21 analysed the symptoms and causes of unsustainability and provided an 'owner's manual' of how to implement sustainable development (Grubb et al. 1993). Building upon the Stockholm Declaration, the Rio Declaration contained 27 guiding principles, including the precautionary principle, equity principles, and the subsidiarity principle (Henns and Nath 2003). The Forest Principles emphasized that while national governments had the right to exploit forests within their borders, they also had a responsibility to ensure that these activities did not negatively affect other countries (Grubb et al. 1993). These documents thus reinforced the importance of collaboration and coordination.

The World Summit on Sustainable Development (WSSD) in Johannesburg focused on issues of implementation since Rio. However, Steiner (2003, 33) believed that an overloaded agenda, poor direction, and a lack of leadership prevented real progress at a formal level. Discussions focused on trade and economic considerations rather than on the merging of these concerns with environmental and social issues. The problems that had inhibited implementation after the Stockholm Conference continued to impede progress.

In the following sections, we will examine the activities of two public agencies synonymous with IREM. The US Forest Service and the Ontario conservation authorities were established in 1905 and 1946, respectively. The histories of these two long-standing and ongoing agencies provide insight into the practice of IREM before it formally arrived on the global stage. They also show how IREM evolved from starting points related to forestry and water management in both American and Canadian contexts.

ONGOING EXPERIENCE IN THE UNITED STATES: THE US FOREST SERVICE

The US Forest Service, located in the US Department of Agriculture, is responsible for managing the 772,953 km^2 of the nation's national forests. In 1891 a forest reserves initiative began, based on existing federally owned land mainly located in the western United States. Gifford Pinchot, the first head of the US Forest Service, 'believed that conservation should be based on three principles: development (using existing resources for the present generation), the prevention of waste and the development of natural resources for the many, not the few' (McCormick 1995, 15). According to McCormick (1995), mainstream resource managers, who were mainly professionals in forestry, agronomy, and hydrology, were very reluctant to remove resources from commercial development because this was against the three principles that formed the core of the 'gospel of efficiency', the dominant resource management philosophy (Hays 1968). Indeed, so strong was their loyalty to their profession's principles that these often overrode the wishes of the public (Cortner and Moote 1999). Initially, forest policy was strongly focused on the use of timber to promote regional and national growth, although local use, including non-timber uses, would be accommodated. This was the essence of US forestry policy for the next 55 years, until the Multiple-Use Sustained-Yield Act of 1960.

From its earliest days, multiple use became a central concept in guiding the activities of the US Forest Service and those of many other public agencies and private interests worldwide. In the context of early US forestry policy, multiple use implied a broad range of purposes, from outdoor recreation, range timber, and watersheds to wildlife and fish, all of which were explicitly recognized in the Multiple-Use Sustained-Yield Act. That statute required that all renewable forest resources be managed to best meet the needs of the American people. Thus, in theory, all uses of the forest were to be given equal consideration in planning and allocation (Fedkiw 1998), requiring the integration of diverse interests and needs. However, in practice, 'best use' was often measured by financial returns, which promoted commercial timber uses.

The Multiple-Use Sustained-Yield Act was the first of several statutes that would affect the operations of the Forest Service. These included the Wilderness Act (1964), the Historic Preservation Act (1968), the National Trails and Wild and Scenic Rivers Acts (1968), the National Environmental Policy Act (1969), the Clean Water Act Amendments (1972), the Endangered Species Act (1973), and the National Forest Management Act (1976). Collectively, these acts necessitated the coordination of a broader range of mandates and directives.

Underlying this legislative change were growing concerns about the definition and practice of multiple-use management. At least two competing perspectives emerged as early as the 1940s. On the one hand, Pearson (1944) believed that multiple use involved a mosaic of uses, dominated by the most preferred or important use. For the US Forest Service, commercial forest production was the primary use, after which other uses might be accommodated; conflicts between the preferred forestry use and other uses would be resolved by separating them in either space or time. On the other hand, Dana (1943) maintained that multiple use referred to several uses occurring at a single place at the same time. This view held that it would be difficult to avoid multiple uses on lands that contained multiple and varied resources.

Initially, forest management focused on ensuring a secure supply of forest products to support economic activity and growth. Between 1905 and 1911, the area under national forests increased by almost 404,686 km^2. Efforts to control forest fires were initiated. By 1908, allowable cut plans were developed for every national forest. In 1924 the Forest Service designated its first forest wilderness area, and by 1945, 8.5 per cent of the national forests had been withdrawn from commercial cutting and placed in wild and wilderness area zoning (Fedkiw 1998). Essentially, the Forest Service had adopted Pearson's view of multiple use—small islands of other uses surrounded by the predominant commercial forestry use. While collectively the multiple-use approach was being embraced, it is unclear how the integration of the range of uses in space and time should or could occur (Fedkiw 1998).

After World War II, changing societal attitudes, growing needs and wants, and improving economic conditions stimulated significant changes within the US Forest Service. At a societal level, there was an increased demand for goods and services from a rapidly expanding and increasingly affluent and mobile society. This trend translated into an increased demand for forest products (especially to support new housing), as well as for space to accommodate outdoor recreational use (Hays 1987). The magnitude and frequency of conflicts between logging and recreation interests increased, stretching the traditional approaches to resolving such disputes.

One shortcoming of the early practice of multiple-use forestry concerned the impacts associated with outcomes of management decisions. For instance, while fire prevention and suppression delivered the desired outcome—a reduced loss of forests from fires, from 20,234 km^2 in 1910 to 2,428 km^2 acres per year by 1920—an associated impact was a rise in fire fuel and increased long-term fire damage potential (Fedkiw 1998). Subsequently, an emerging scientific awareness of the importance of fires in the life cycle of many forest ecosystems prompted the need for prescribed burns. This activity, perceived as contradicting fire suppression efforts, was not easily accepted in some quarters. In wildlife and wilderness areas, animal populations sometimes increased beyond an area's carrying capacity, while on the other hand, the plight of the northern spotted owl owing to logging operations in the Pacific Northwest gained widespread public attention in the mid-1980s. The numbers of people making recreational demands increased from one million in 1910 to 236 million in 1981, leading to concern about the impacts of these users and how they might be managed (Fedkiw 1998).

In the 1960s, other public concerns arose. Although forestry plans identified allowable cuts, overharvesting of forests had occurred. Pork barrel politics often seemed to be operating very well between the Forest Service and Congress (Cortner and Moote 1999). Calls arose for better managers who would reflect broader interests. A subsequent review of forest landholdings would show that the area allocated to timber was reduced from 75 million in 1969 to 53 million acres in 1993. The practice of clear-cut forestry was significantly reined in. The 1970 National Environmental Policy Act allowed the public to challenge Forest Service management practices such as clear-cutting. Public participation in the management of forests became the rule. Preserving scenic landscapes from forest and mining activities became common practice. In 1995 Congress prohibited the sale of timber products at below market price (Fedkiw 1998). All of these changes allowed more interests to be considered, and this in turn meant that there had to be more collaboration and coordination—and, ultimately, integration.

In 1992 the Forest Service announced a new philosophy—ecosystem management. Salwasser (1997) explained that, in the context of forests, ecosystem management integrated social, economic, and environmental goals at different spatial and temporal scales, and maintained the diversity of forest life forms, ecological processes, and human cultures. This interpretation, however, fails to clarify the purpose of ecosystem management. Is the purpose to sustain current and future uses that meet people's needs, or is it to maintain the forests' biophysical processes, or is it some combination of these two purposes? Before an integrated approach can be applied, the ultimate direction or vision has to be clear.

ONGOING EXPERIENCE IN CANADA: ONTARIO CONSERVATION AUTHORITIES

The Ontario conservation authorities were established to facilitate coordination among municipal governments, as well as between municipal and provincial governments, and to promote a comprehensive approach to resource management. Specifically, the conservation authorities were mandated 'to further the conservation, restoration, development and management of natural resources, other than gas, oil, coal and minerals.' In 1941 a conference was held in Guelph to consider how water and land management could become more effective. At that time, Canada was involved in World War II, and although that war would continue for many years, senior governments were thinking about how jobs could be provided for returning war veterans. A different concern was the degraded state of the renewable natural resources of Ontario, on which the province's agricultural and forest-based economy relied. The conference petitioned senior governments to undertake a study to identify what needed to be done and what the potential benefits would be. The Ganaraska River watershed was selected for the study. With respect to the implementation of any conservation program, the report, titled *The Ganaraska Survey*, concluded that 'the use of an existing provincial agency would be the most obvious solution . . . but . . . no one department is equipped at present with a staff of experts trained in all sciences represented. Even if a department were so equipped, it is questionable whether the best interests of the community would be served by having a government department take absolute responsibility for such a programme' (Richardson 1944, 238). 'River valley development' was to be the approach taken, through new watershed-based organizations (called conservation authorities) based on the following principles:

1. *Watershed as the management unit.* Many of the economic staples of the province, such as agriculture and timber, depended on water and terrestrial resources, highlighting the need for a functionally integrated approach. The provincial government had examined the Conservancy Districts in Ohio (Jenkins 1976) and the Tennessee Valley Authority (Hodge 1938; Ransmeier 1942) and decided that a catchment-based organization would be the best way to ensure that interrelationships among various resources and human interests were addressed.

2. *Local initiative to be essential.* A conservation authority would only be established when two or more municipalities in a watershed agreed they wanted to collaborate. On that basis, they could approach the provincial government to establish a conservation authority. Once municipalities agreed to form one, they became eligible for provincial funding not available to local governments operating individually. In this manner, a powerful incentive was provided for municipal governments to become involved in a conservation authority.

3. *Provincial-municipal partnership to be a core aspect.* Although the provincial government would not impose a conservation authority, it would participate as a partner. As Lord (1974, x) observed, 'Cost sharing resolved itself into 50 per cent by the authority and 50 per cent by the government of Ontario. This proved to be one of

the soundest ideas of the authority movement. It has meant that an authority can flourish only when the local people have enough enthusiasm and conviction to support it financially. It has also meant that the authority does not exceed the financial resources of its jurisdiction.' However, this feature also meant that areas with few people or a modest tax base would not be able to form a conservation authority, because there would not be the local capacity to raise the required funds.

4. *Comprehensive perspective required.* Many land-based problems were caused by too much or too little water, and water-based problems often were influenced by land-based activities. Thus, a 'comprehensive approach' was promoted, meaning that water and associated land-based resources would be considered together. This would be facilitated by using a river basin or watershed as the management unit, as it was appreciated that activities in or decisions taken for one part of a catchment had implications for other areas in the catchment.

5. *Coordination and cooperation to be pursued.* Any new conservation authority was required to create links with provincial and municipal agencies responsible for other natural resources, the environment, and planning. For a project focused on land use, the conservation authority was required to seek assistance from the provincial Department of Agriculture, or for a project focused on forestry, to seek help from the Department of Lands and Forests. Between 1946 and 1953, such collaboration and cooperation were institutionalized with a stipulation for multi-departmental approval of projects undertaken by conservation authorities. Such top-down overseeing was complemented by a bottom-up approach in that conservation authorities could create advisory boards for aspects of their programs. Membership on advisory boards was open to people from the public and provincial agencies (Mitchell and Shrubsole 1992).

The Ganaraska Survey served as a model for subsequent planning documents, called Conservation Reports, that were prepared by the provincial government when a conservation authority was formed. Multiple-purpose use, one principle of river valley development, was incorporated into the management of forests. As with the experience of the US Forest Service, the achievement of multiple use by conservation authorities has been a challenge for several reasons. First, increased competition for the use of the forests has reflected greater affluence, mobility, and technological development. Relatively new users in some forests include snowmobilers, cross-country skiers, motorcyclists, off-road cyclists, and horseback riders. Second, public expectations concerning the state of the natural environment have risen.

As mentioned earlier, coordination and cooperation were important features of the conservation authorities. An important mechanism to achieve these was the advisory board system and the membership on these boards. Membership was not restricted to authority representatives, and it provided an opportunity to obtain information, advice, and assistance from all groups with an interest in conservation (Richardson 1960). Until the 1960s, the employees of provincial agencies—especially Forestry and Agriculture—were full voting members. This meant that these individuals, in addition to implementing the policies of their closely related provincial agencies, had influence over the policies adopted by each individual authority. In the early 1960s, it was deemed inappropriate for provincial employees from resource management agencies to serve on advisory boards.

Water management, particularly flood control, was central to the conservation authority program. The authorities have had problems pursuing this objective while serving other societal needs, as illustrated in the Grand River valley. The Grand River Conservation Authority is the largest of the 33 conservation authorities located in southern Ontario, with an area of about 7,000 km^2. Rural land uses are located north and south of urban settlements generally concentrated in the central portion of the catchment. The *Grand River Hydraulics Report* of 1954 provided a water management vision for the Grand River Conservation Authority that endured until the 1970s. Its purpose was to ensure understanding of the hydraulic nature of the Grand River system and to identify how the river could be 'controlled'. 'Control deals with the prevention of floods by the use of reservoirs and other structures, and the increase of summer flows' (Ontario 1954, 1). The report recommended some upstream dams and channel improvements in several communities, at a total cost of $22,110,000 (Cdn). Two reservoirs were completed, one in 1958 and the other in 1974. Various cost-benefit analyses were completed in the 1960s in support of the construction of the other reservoirs. However, senior governments did not approve financial assistance because it was unclear to what extent these and other initiatives, including a pipeline network from the Great Lakes to provide water to communities in the Grand River valley and beyond, best served public needs. In addition, by the late 1960s, public opposition to the construction of some of the dams had become significant. This shift was consistent with the general trends noted earlier in conjunction with the Stockholm Conference.

Part of the problem was the limited frame of reference used by the conservation authority and other public water management agencies. Each agency usually made plans on the basis of its limited legislative mandate (MacIver 1970). Indeed, water management was a competitive rather than a collaborative process. It was not until 1978 that an inter-agency and inter-municipal strategic planning effort was undertaken in the Grand River valley. After an expenditure of $1.6 million (Cdn), a report was published in 1982 that integrated concerns about water supply, water quality, and flood damage reduction. Detailed technical studies were completed by five subcommittees and several advisory groups. Public participation was a central component, a departure from the technical approach that had dominated watershed planning in the 1950s and 1960s. Twenty-six alternative plans were evaluated with respect to their ability to fulfil water objectives, cost, social and environmental impacts, and public acceptability. A range of options, including structural and non-structural alternatives to flooding, water supply, and sewage treatment, was considered. Although the provincial government did not officially endorse the recommendations in the report, A.F. Smith, the study coordinator, later observed that 'tacit unofficial approval was given since the major portions of the plan were implemented without formal endorsement.'

In 1994 the Grand River was designated as a Canadian Heritage River on the basis of its recreational and heritage values (Nelson et al. 2004). Spearheaded by the Grand River Conservation Authority, the heritage river initiative represented a significant departure from past activities. Regarding IREM, the frame of reference extended well beyond conventional water and related land resources (e.g., wetlands) to include and integrate human heritage and recreational values. This point is illustrated in the words of Barbara Veale (2004, vi–vii), chair of the Heritage River Committee:

> Surface and groundwater areas are protected from encroachment and contamination. Surface and groundwater within the watershed provide sufficient potable water to support sustainable urban and rural growth.... Pedestrian and bicycle trails make use of natural areas to link residential, commercial and industrial areas to river corridors. New residential subdivisions are compact, energy efficient and designed to harmonize with natural and cultural heritage features.... Industries have adopted new processes to eliminate harmful air emissions and waste by-products and to reduce energy and water consumption.... The cultural landscapes shaped by early aboriginal and European settlements are still evident. Innovative farming techniques, conservation measures and technological advances have increased agricultural productivity while reducing chemical and organic runoff into local creeks and streams. Natural corridors and forests have rejuvenated and expanded. They are now connected throughout the rural countryside, providing wildlife habitat, representative flora and fauna, vegetative buffers and renewable timber.

From an IREM perspective, progress has been made to manage surface and groundwater resources together, particularly for the protection of source waters. This is, in part, a response to the tragedy that occurred in Walkerton, Ontario, where seven people died and over 2,000 became ill as a result of contamination in the town's groundwater supply in May 2000. A subsequent inquiry led to recommendations pointing to the urgent need for an integrated approach to source area protection. One tangible outcome from the inquiry was the passage of the Safe Drinking Water Act in 2002, which called for protection of water from source to tap. Given their founding principles, conservation authorities are well placed to undertake their designated responsibilities for source water protection.

In Ontario, and in the Grand River in particular, IREM has evolved from an initial reliance on structural adjustments that provided for multiple uses through single means (e.g., dams and reservoirs) to a reliance on both structural and non-structural adjustments. A wider consideration of water management activities was promoted through the 1980s. Public participation and stakeholder involvement have emerged as crucial elements in achieving effective strategic planning and program implementation. The 1990s saw water management initiatives further broadened through attention to heritage issues. Currently, concerted efforts strive to integrate all aspects of the hydrologic cycle—surface, ground, and air—into management plans and technical scenarios.

LESSONS

Based on analysis of experiences across Canada, in the Great Lakes Basin, and in other nations, Mitchell (1998) identified 'lessons learned' related to design and implementation of IREM, especially for aquatic systems. No lesson by itself is sufficient to ensure effective application of an integrated approach. However, all of them represent accumulated insight about IREM:

1. Importance of understanding and appreciating the *context* or local conditions that require the capacity to custom-design solutions.
2. Appreciation of the need to take a *long-term perspective*, since resource and environmental problems were usually not created in a few years and therefore are unlikely to be resolved quickly.
3. Importance of having a *vision* or *direction* so that there is a clear understanding of what desired future condition is sought.
4. *Legitimacy* or *credibility* must be established, best achieved through *ongoing commitment* from political and other leaders in local communities.
5. A key success factor is to have a *leader* or *champion* who will continue to work for and support an integrated approach through inevitable setbacks, disappointments, and frustrations. Such a committed leader is often the *key* success factor.
6. A willingness to *share* or *redistribute power* is usually necessary if significant change is to occur.
7. A *multi-stakeholder group* should be created to ensure that the IREM process is representative, open, transparent, and accessible.
8. Decisions by the multi-stakeholder group should be based on *consensus*, which is the best way to ensure the community's long-term commitment to accept decisions.
9. Volunteers who participate in the management process should be aware of the likelihood of *burnout*.
10. It should be accepted from the beginning that a process for IREM will normally unfold in conditions characterized by *turbulence, uncertainty*, and *surprise*. As a result, participants need to be flexible, adaptable, and able to learn from experience.
11. There never can be enough time devoted to *communication* within and between groups. Such communication should be done in 'plain' language to keep all participants informed and updated.
12. *Demonstration projects* should provide tangible evidence of progress and allow a role for those who feel more comfortable with a hands-on rather than a planning approach.
13. Long-term change and improvement require attention to *information* and *education*, since such change is most likely to occur from different attitudes and values in the community.
14. Explicit provision should be given to means for *implementation* and *monitoring* of progress regarding decisions and plan recommendations.
15. *Accomplishments* should be noted and celebrated.

LOOKING FORWARD

Defining an Integrated Approach

If integrated resource and environmental management is to contribute to effective, efficient, and equitable achievement of a vision, there must be a clear understanding about what the concept means. Otherwise, different expectations may be created, and it will be difficult to identify expected results (outputs, outcomes, and impacts).

One danger is that a 'comprehensive' interpretation may be taken. This refers to planning efforts that attempt to understand all aspects of the natural and human environment and their interrelationships. One of the early criticisms of the comprehensive approach to river basin planning in Canada was that it took too long to complete (Cardy 1981; O'Riordan 1981; Mitchell and Gardner 1983; Mitchell 1983). To streamline and focus the process, it was later suggested that we should focus on selected variables and relationships, especially those that account for the main variation in system behaviour and are amenable to management intervention. In that regard, an integrated approach maintains the core principle of systems analysis by recognizing the importance of considering entire systems, along with their variables and relationships, but it is pragmatic in that it focuses on only selected variables and relationships.

Inevitable Tension

Collaboration and cooperation are essential for successful integration. These are usually achieved through partnerships, not only among government agencies, but also with the private sector and civil society. However, tension often arises when integrated and partnership approaches are combined. An integrated approach, while focused and selective, still directs attention to the entire system, its parts and relationships. This requires a big-picture perspective. In contrast, partnership approaches, when drawing upon local people, can result in a focus on local issues, problems, and opportunities, as these are what local people understand best. This can lead to a small-picture perspective and lack of attention to or concern about the overall region or system.

As a result, when combining integrated and participatory approaches, both of which are desirable and appropriate, it is important to be sensitive to this possible tension. One way of overcoming this challenge is to work at two scales simultaneously. For example, in a river basin or region, it has been common to create a number of local stakeholder groups representing different subcatchments or subregions. Each of these groups focuses on issues within its area. However, there is also a group that addresses 'whole of basin or region' issues. The latter group's membership includes individuals from each of the sub-basin or subregional groups. In this way, it is possible not only to get insights and benefits of local understanding related to localized issues, but also to use that to inform thinking about the entire river basin or region.

Alternative Organizational Approaches

IREM can be implemented by a single multiple-purpose agency, as well as by more specialized agencies in combination with some coordinating mechanism. Every approach has strengths and weaknesses. However, it must always be remembered that, in attempts to align management functions with organizational structures, perfect fits or matches are rare. Furthermore, what are known as *edge* or *boundary problems*, or situations characterized by shared or overlapping interest, responsibility, or authority, will be normal. Such edge problems are not resolved by redesigning organizational structures. Indeed, in most cases, bothersome *inter*departmental problems become *intra*-departmental issues when a number of agencies are combined into one larger organization. Thus, it is not possible to remove or eliminate edge or boundary problems when many interests are involved. The 'edges' must be recognized, and processes and mechanisms created so that they can be managed.

Voluntary or Prescriptive Strategies

There are ways to help agencies, organizations, groups, and individuals collaborate, coordinate, and integrate activities. What might be called a *voluntary option* can be effective when there is goodwill, trust, respect, and willingness to work together (Hooper et al. 1999, 750). In this type of situation, emphasis is upon encouraging consideration of each other's goals and processes, with no specification of performance goals or outcomes. Instead, attention is directed towards benefits to be realized through integration, collaboration, and coordination. Existing agencies are used, with the addition of a mutually acceptable coordinating mechanism.

At the other end of the spectrum is a *coercive option*. Here, compliance may be a problem because of lack of willingness to work together. Thus, prescription is viewed as necessary, and actions and processes are stipulated. A new agency is established and given some or all of the responsibilities and authority previously held by other agencies. The overall intent is to induce adherence to policy prescriptions and regulatory standards.

An intermediate position is a *cooperative option*. The background here is a situation in which compliance is not judged to be a problem and discretion is encouraged in the development of policy and coordination. Goals and processes, as well as expected outcomes, are prescribed. No new integrating agency is established, but one or more existing agency is designated to have a lead role. Different agencies may have lead roles, depending upon the issue or problem being addressed.

These options are presented as distinct, occupying different places along a continuum. However, hybrids can be developed. For example, a coercive option may be used initially in a situation in which there seems to be little will to work together. As time passes and participants demonstrate the capacity to work on a collaborative basis, adjustments may be made to move participants towards the cooperative option. Alternatively, a minimalist approach might be to begin with a voluntary option and only move to the more prescriptive options if the participants do not demonstrate the goodwill, trust, and respect necessary for the voluntary option to work well.

The answer to the question about how IREM should be implemented lies in the realm of public policy. What is required is the engagement of academic, public policy, and civil society groups in healthy debate over the sometimes fuzzy nature of the concept. In this way, the lessons emerging from the successes and weaknesses of IREM in practice can be identified and then used to improve its future implementation.

REFERENCES AND FURTHER READING

Cardy, W.F.G. 1981. River basins and water management in New Brunswick. *Canadian Water Resources Journal* 6 (4): 66–79.

Cortner, H.J., and M.A. Moote. *The politics of ecosystem management.* Washington, DC: Island Press.

Dana, S.T. 1943. Editorial: Multiple use, biology and economics. *Journal of Forestry* 41 (9): 625–6.

Fedkiw, J. 1998. *Managing multiple uses of national forests: A 90-year learning experience and it isn't finished yet.* Washington, DC: US Department of Agriculture, Forest Service.

Grubb, M., M. Koch, A. Munson, F. Sullivan, and K. Thomson. 1993. *The 'Earth Summit' Agreement: A guide and assessment.* London: Earthscan Publications.

Hays, S.P. 1959. *Conservation and the gospel of efficiency: The progressive conservation movement, 1890–1920.* Cambridge, MA: Harvard University Press.

——— 1987. *Beauty, health and permanence: Environmental politics in the United States, 1955–1985.* New York: Cambridge University Press.

Henns, L., and B. Nath. 2003. The Johannesburg Conference. *Environment, Development and Sustainability* 5:7–39.

Hodge, C.L. 1938. *The Tennessee Valley Authority: A national experiment in regionalism.* Washington, DC: American University Press.

Hooper, B.P., G.T. McDonald, and B. Mitchell. 1999. Facilitating integrated resource and environmental management: Australian and Canadian perspectives. *Journal of Environmental Planning and Management* 42:747–66.

Jenkins, H.A. 1976. *A valley renewed: The history of the Muskingum Watershed Conservancy District.* Kent, OH: Kent State University Press.

Lord, G.R. 1974. Introduction to A.H. Richardson (ed.), *Conservation by the people*, ix–xi. Toronto: University of Toronto Press.

McCormick, J. 1995. *The global environmental movement.* 2nd edn. Toronto: John Wiley & Sons.

MacIver, I. 1970. *Urban water supply alternatives: Perceptions and choice in the Grand Basin, Ontario.* Research Paper No. 126. Chicago: Department of Geography, University of Chicago.

Mitchell, B. 1983. Comprehensive river basin planning in Canada: Problems and opportunities. *Water International* 8:146–53.

——— 1986. The evolution of integrated resource management. In R. Lang (ed.), *Integrated approaches to resource planning and management*, 13–26. Calgary: University of Calgary Press.

——— 1990. Integrated water management. In B. Mitchell (ed.), *Integrated water management: International experiences and perspectives*, 1–21. London: Belhaven.

——— 1998. *Sustainability: A search for balance.* First Annual Research Lecture, Faculty of Environmental Studies. Waterloo, ON: University of Waterloo.

Mitchell, B., and J.S. Gardner (eds). 1983. *River basin management: Canadian experiences.* Department of Geography Publication Series No. 20. Waterloo, ON: University of Waterloo.

Mitchell, B., and D. Shrubsole. 1992. *Ontario Conservation Authorities: Myth and reality.* Department of Geography Publication Series No. 35. Waterloo, ON: University of Waterloo.

Nelson, J.G., with B. Veale, B. Dempster, et al. 2004. *Towards a grand sense of place.* Waterloo, ON: University of Waterloo, Heritage Resources Centre.

Ontario, Department of Planning and Development. 1954. *Grand River hydraulics report.* Toronto: Ontario Department of Planning and Development.

O'Riordan, J. 1981 New strategies for water resources planning in British Columbia. *Canadian Water Resources Journal* 6 (4): 13–43.

Pearson, G.A. 1944. Multiple use forestry. *Journal of Forestry* 42 (4): 243–9.

Ransmeier, J.S. 1942. *The Tennessee Valley Authority: A case study in the economics of multiple purpose stream planning.* Nashville: Vanderbildt University Press.

Richardson, A.H. 1944. *The Ganaraska Watershed.* Toronto: King's Printer.

——— 1960. Ontario's conservation authority program. *Journal of Soil and Water Conservation* 15 (5): 252–6.

Salwasser H. 1997. Multiple use, ecosystems and diversity. In V.A. Sample, R. Weyerhaeuset, and J.W. Giltmier (eds), *Evolving toward sustainable forestry: Assessing change in U.S. forestry organizations*, 210–11. Washington, DC: Pinchot Institute for Conservation.

Schramm, G. 1980. Integrated river basin planning in a holistic universe. *Natural Resources Journal* 20:787–806.

Steiner, M. 2003. NGO reflections on the World Summit: Rio + 10 or Rio – 10. *RECIEL* 12 (1): 33–8.

Veale, B.J. 2004. *The Grand River, Ontario: A decade in the Canadian Heritage Rivers System: A review of the Grand Strategy 1994–2004.* Cambridge, ON: Grand River Conservation Authority.

Weber, E.W. 1964. Comprehensive river basin: Development of a concept. *Journal of Soil and Water Conservation* 19 (4): 133–8.

White, G.F. 1957. A perspective on river basin management. *Law and Contemporary Problems* 22:157–87.

World Commission on Environment and Development. 1987. *Our common future.* London: Oxford University Press.

CHAPTER 3

Research and the Integration Imperative

Stephen Dovers and Richard Price

1. INTRODUCTION

This book presents recent thinking and practice in integrated resource management and evidences the strength of the *integration imperative*. The development of integrative resource management regimes requires new forms of information and knowledge and thus new forms of research. This chapter explores why integrated research is needed and how it can be pursued.[1] While research and development (R&D) is the focus, the chapter also addresses integrated policy and management. To be useful, research for resource management must connect to activities outside research organizations, in policy and management agencies, interest groups, and the community.

The chapter proceeds as follows. Section 2 discusses the emergence of the integration imperative, different motivations and understandings, and the challenges presented for research and policy. Section 3 defines the dimensions of integration—the purposes, approaches, and potential contributions. These sections are general in scope, whereas section 4 draws on a case study to explore practical aspects of implementing integrated research, and does so in an intentionally first-hand and active manner. Section 5 proposes a set of principles for integrative and interdisciplinary research.

2. THE INTEGRATION IMPERATIVE

The following quotes reflect an increasingly perceived imperative for integration. They also reflect the typically broad nature of statements of this imperative and the lack of detailed instruction. (Emphasis has been added.)

> 35.9 The scientific and technological means include the following: (a) Supporting *new scientific research programmes, including their socio-economic and human aspects*, at the community, national, regional and global levels, to complement and encourage synergies between traditional and conventional scientific knowledge and practices and strengthening interdisciplinary research related to environmental degradation and rehabilitation. (UN 1992)

103. Improve policy and decision making at all levels through, inter alia, *improved collaboration between natural and social scientists, and between scientists and policy makers,* including through urgent actions at all levels to:
(a) Increase the use of scientific and technological knowledge, and *increase the beneficial use of local and Indigenous knowledge* in a manner respectful of the holders of that knowledge and consistent with national law;
(b) Make greater use of *integrated scientific assessments,* risk assessments *and interdisciplinary and intersectoral approaches.* . . .
(e) Establish partnerships between scientific, public and private institutions . . . by *integrating scientists' advice into decision-making bodies* in order to ensure a greater role for science, technology development and engineering sectors. (World Summit on Sustainable Development: Plan of Implementation, 4 September 2002)

The imperative for integration stems from recognition of the interdependence of human and natural systems, central to the modern agenda of sustainable development. Within resource management, there are longer-standing realizations: for example, land degradation demands recognition and the coordinated management of ecological, climatic, economic, cultural, and institutional elements.

International and national policy and law state the *policy integration principle*: environmental, social, and economic considerations must be integrated rather than dealt with piecemeal to advance the social goal of an ecologically sustainable, socially desirable, and economically viable future (e.g., the 1992 Rio Declaration and Agenda 21, the 2002 Johannesburg Declaration). This three-way integration defines the intellectual and methodological challenge of developing integrative methods and capacity. The intellectual challenge is extended by the fact that integrative capacity demands a sophisticated understanding of interactions among complex, interdependent, and often non-linear human and natural systems. Thus, integration has at least two meanings: *integration in research,* combining multiple disciplinary perspectives; and *integration in policy-making,* connecting agencies, issues, and sectors.

Beyond policy and research, there is also the need for integration of the social, ecological, and economic factors in practical resource management. Research may inform management and policy settings may constrain or enable it, but the variability of management contexts, the needs of land managers (farmers, water catchment officers, foresters, reserve managers, etc.), and the expertise and knowledge they possess indicate that on-ground management is a further domain of integration. Although integration is widely advocated, the way to go about achieving it is not well understood in either the research or policy arena. At this point we can note a delineation in the intent of integration (this will be expanded upon later): integration that is intended only to be *informative* (to inform understanding or to develop policy support tools) or that is aimed at being *decisive* (to formulate integrated policy or management decisions).

2.1. Different Integration Imperatives

The policy integration principle is only one imperative that drives integration. To expose others, we can consider the characteristics of resource management problems that suggest the requirement for more than one policy sector/agency or one disciplinary response. The problem attributes that render sustainability problems different and difficult suggest multiple drivers for integration (drawing on Dovers 1997):

- *Integration in space*, demanded by natural processes operating over variable and extended spatial scales (e.g., whole catchment systems, landscape-wide ecological functions, nutrient cycles). Resource management problems also traverse political, legal, and administrative boundaries, requiring at least coordination if not renegotiation of boundaries.
- *Integration in time*, to address extended and variable temporal scales (e.g., climate, evolutionary processes, non-degradable wastes, species population viability, etc., versus political or economic time scales), and the often cumulative rather than discrete nature of environmental impacts.
- *Integration within and across academic disciplines, professional domains, and policy/management sectors*, demanded by connectivity between substantive issues (e.g., salinity, water quality, and vegetation, or catchment management and fisheries) and between policy sectors (e.g., public health, environmental protection, industry policy, etc.).
- *Integration vertically in social systems and within and across sectors*, to address systemic causes of environmental degradation deep within patterns of production and consumption, settlement and governance. This demands attention to indirect causes of problems rather than the simple treatment of symptoms (or corrective versus antidotal measures, Boyden 1987).
- *Integration of understanding of natural systems, economic drivers, legal and institutional contexts, and social and psychological factors in policy design and implementation*, driven by the need for innovative policy and management approaches.
- *Integration of different segments of society and their knowledge systems* (firms, governments, civil society, Indigenous cultures, research institutions), in view of the need to involve many communities in natural resource management and to establish partnerships between private, public, and community interests.
- *Communication as integration*, where transfer, wider ownership, and uptake of existing, possibly unconnected information can serve to advance integration.

The last two imperatives provide a further arena of integration: *integration through participation*. This is closely related to the other three (research, policy, and management) and is no less problematic or imperfect in practice. It also contains different aspects: integration of community into management programs; integration of non-government players into higher-order policy processes; and integration of formal (disciplinary-based) and community or traditional knowledge. The claim to validity through a wider range of knowledge systems is a feature of contemporary sustainability debates. All this suggests different but related aspects of integration, and thus the importance of defining the *purpose* of integration—identifying which of the above problem characteristics are most relevant before selecting the appropriate methods or processes.

2.2. When and When Not to Integrate

If integration is required, then 'dis-integrated' or singular research and policy approaches must have been tried, evaluated, and found wanting. If a single-sector policy or single-discipline R&D approach is available but is not, or is insufficiently, used, attempting the harder task of integration could be an inefficient diversion. *Integration* and *interdisciplinarity* are terms in danger of becoming mere passwords in workshops and funding applications. To counter this and ensure rigour and best use of resources, the *problem definition* phase in either policy or research needs to be emphasized and the need for integration justified by reference to the lack or failure of non-integrated alternatives.

Flowing from this consideration is the recognition of *degrees of integration*. Full integration of environmental, social, and economic considerations and the many relevant disciplines is not necessary in every case. Partial integration may sometimes be appropriate. Additive (not integrative) *multi*-disciplinary R&D, or the partial integration of community representation into a policy process, may be sufficient, or input by a social scientist into problem definition and research design in a biophysical science project (or by a biophysical scientist into a social research project) may suffice at early phases but not so much in subsequent phases. Similarly, for some policy challenges, modest involvement of another portfolio or agency may be sufficient. In other cases, more substantial and sustained integration may be required and may demand substantial institutional reform.

2.3. Motives and Interests in Integration

Although the academic literature and environmental activists recognized the need for integration some time ago, the situation now is not so clear. While new, interdisciplinary research programs are the major loci of activity, equally strong calls come from policy and management agencies and from community stakeholders in resource management who confront inseparable ecological, social, and economic issues. While all these parties may agree that integration is required, what they mean by integration and what fires their interest will vary. The example of integrated catchment management (ICM) provides an illustration. ICM aims to integrate aspects of resource management that have been managed through separate agencies, policies, and processes. ICM has widespread currency, has been widely implemented, and is an approach still under construction. The players and motivations in ICM can be characterized as follows:

- Natural scientists who understand one aspect of the biophysical system (e.g., landscape ecology, geomorphology, hydrology) perceive interconnections in resource management problems and seek to integrate with other disciplines. Aquatic ecologists and hydrologists might address riparian and stream biodiversity as affected by flow regimes, while ecologists might work with economists to establish methods for ecosystem valuation.

- Social scientists from various disciplines, driven by intellectual curiosity and research funding, seek ways of incorporating social, cultural, political, legal, and economic dimensions into resource management, a field they perceive to be dominated by natural science and administrative rationalities. Rural sociologists and environmental psychologists seek to understand landholder understandings of biophysical issues and to integrate these with policy-making. Black-letter lawyers explore issues of non-compliance, while law-in-context researchers examine regulatory implementation at finer scales. Economists advance game theory and agent-based modelling to understand land managers' behaviours, multi-criteria analysis to integrate values, and tradable rights to serve as policy instruments.[2]
- National policy-makers, seeking efficiency and effectiveness in program delivery and expenditure of government funds, pursue the development of a generic approach for integrating salinity, water quality, and biodiversity, to be implemented at regional or catchment scale through accredited plans.
- State/provincial and local governments seek to balance resource management objectives with regional development and employment, development of agricultural industries, and maintenance of downstream town water quality, emphasizing trade-offs between these imperatives through negotiation processes and coordination between agencies.
- Rural landholders, focused on farm viability and maintenance of the natural resource base, concentrate on involvement, at district scale, of community-based groups engaged in on-ground management activities and on financial and taxation assistance to encourage conservation works. They emphasize the need for integrated frameworks for data gathering and for recommending land-use options.
- A research funding agency is tasked with investing in integrative R&D to produce operational strategies, where proposed integrative approaches are contestable and demand exceeds the supply of funding and where partnership arrangements with other agencies are difficult to develop.

These variations are defined by a limited range of issues (water quality and quantity, salinity) at a particular scale (the catchment). Other issues relevant to integrated resource management are not suited to research or policy at catchment scale (infrastructure, communications, education, community development, industry policy, public health, etc.). Moreover, the weak institutional and statutory base of catchment management in most jurisdictions creates problems in maintaining integrative initiatives and linking catchment management with other policy sectors.

2.4. The Magnitude of the Integration Task

While there is overlap among these interpretations of integration, the motives behind them, and the collaborative directions they invite, there are also differences to negotiate among a range of players. The ICM context is relatively well understood, but the preceding list indicates that integration in ICM is complex, difficult, and full of contested methods and proposed institutional structures. If we widen the scope to all possible contexts of integration—issues other than catchment management, whole-of-government institutional reform, cross-sectoral policy assessment, global issues, and so on—the task is more difficult still.

Integrated resource management will be significantly more difficult—intellectually and practically—than non-integrated research, policy, and management approaches. This shift comes with the modern idea of *sustainability*: we have to do both more and better. If the implications of this for intellectual, human, financial, and institutional resources are ignored, failures in R&D and policy are inevitable (see Connor and Dovers 2004). Although there have been significant advances in theory and method—and to some extent in policy and institutional reform—for integration, the magnitude of the task should be appreciated. Sustainability is a higher-order social goal akin to democracy, justice, or equity. Such goals are long term, pervasive, and always contestable in definition and implementation. Over the next few decades, it can be expected that the integration imperative will be continuously addressed and advances made. But it will take time and coordinated efforts, and there will never be a singular approach.

Widely accepted integrative metrics capable of guiding decisions will never exist. At best, integrative R&D and policy processes can better inform decisions, but not make them. Integrative approaches will not make political choices redundant, and the contribution of integrative methods to policy and management is to identify and describe connections and inform more sophisticated trade-offs. Only a political decision can incorporate incommensurable information and values.

The above identifies but does not define the many dimensions of integration. It stresses the importance of a framework within which different integrative initiatives can be viewed to allow understanding and negotiation of these initiatives and to inform choices in a complex area.

3. DIMENSIONS OF INTEGRATION

This section presents a categorization of the five dimensions of the integration imperative: why integrate (purpose); how to integrate (methods); participation as integrative strategy; issues of scale; and expertise and integration. Comment is also made on communication and learning.

3.1. Why Integrate: Definition of Problem and Purpose

The impetus for integrated research arises from two requirements for sustainability: the integration of ecological, social, and economic factors; and the integration of different interests. These two requirements inform the following categorization of the purposes of integration, as does a differentiation between integration that is aimed primarily at *informing* policy and management and integration that is *decisive* in producing policy options.

1. *Integration of ecological, social, and economic factors*, accepting that integrated resource management or sustainable development cannot be significantly advanced while these are considered separately. This has two aspects:

(a) To increase understanding of linked phenomena—that is, of interdependencies within and between natural and human systems—whether as desirable knowledge in itself or driven by a defined policy problem. This involves the integration of existing separate knowledge and understanding and newly created bodies of knowledge.
(b) To inform the design of policy processes, organizational structures, and institutional settings to enhance capacities to integrate environmental, social, and economic factors.

2. *Implementation of integrated policy and management*, through prescriptions for policy instruments and processes, institutional reform, or management interventions. Such prescriptions may emerge from (1) above, and in publicly funded R&D, there may be sensitivities over research that prescribes policy or institutional options.

3. *Integration of differing interests* through community participation and stakeholder involvement in research, policy, and management. This reflects the belief that policy and management will therefore be more effective and will approach the ideal of participatory politics.

Although simplified, this delineation is a first step in understanding some of the complexity of the task, and it allows further specifications on *how* and *who*, undertaken below.

3.2. How to Integrate: Methods and Processes

There are multiple means of addressing these imperatives for integration, and five broad categories are identified below: interdisciplinary research, methodological development, applied problem-solving, policy and institutional design, and communication-as-integration.

Interdisciplinary and Multidisciplinary Research and Development

Research that combines multiple disciplines is key to understanding linked phenomena and to informing policy and management settings. A history of increasing disciplinary specializations, each with its own 'epistemological commitments' (Schoenberger 2001), theories, methods, data requirements, and so on, has allowed understandings at finer resolutions but can work against integrated understanding. Single-discipline research remains crucial to knowledge generation for resource management, as is *additive* multidisciplinary R&D when it does not question the operating assumptions or methods of contributing disciplines. Similarly crucial is largely single-discipline work that incorporates knowledge from another discipline without questioning its source or validity.

Most difficult is interdisciplinary research with *transformative potential* for the participating disciplines. If sustainability problems are different and difficult enough to present serious challenges to existing understanding and policy approaches (Dryzek 1987; Dovers 1997), then there is a *prima facie* case that disciplines and related professional domains, out of which such understanding and approaches have developed, may be deficient. Interdisciplinary research demands the exposure and questioning of theory and method, which in turn require a critical or reflexive capacity within the research process. Such questioning has been a feature of some interdisciplinary activities to date, such as in ecological economics (Common 2003; Dovers et al. 2003). Assumptions in neo-classical economics regarding rational utility maximizing behaviour and consumer sovereignty have received sustained critical attention. Other disciplines and their assumptions (e.g., ecology, law, public policy) have received less scrutiny. The need for transparency and critical evaluation instructs that the integration of disciplinary perspectives cannot be an add-on but must be central to problem definition and research design.

The contribution of different disciplines depends on the problem at hand. Some disciplines have been prominent in integrative endeavours in areas such as economics, rural sociology, hydrology, and ecology. Others have been less prominent. The optimal combination of disciplines cannot be prescribed without reference to the key variables and processes operating in a given context. Similarly, the appropriate style and degree of integration will vary. It may be sufficient for, say, a lawyer or economist to be briefly involved in the problem-framing and research design of a biophysical project so as to ensure relevance to the regulatory or incentive setting. Alternatively, sustained involvement of a larger number of disciplines may be required, generating new theoretical propositions, methods, and insights.

Some disciplines have a history of interaction and work together more easily. Some, such as ecology, geography, and public policy, are already methodologically and theoretically diverse. Natural scientists might connect more easily with other scientists than with qualitative social scientists, and vice versa. Connections across major disciplinary divides—social and natural sciences, the humanities—might be expected to be more difficult to achieve. But it is across those divides that sustainability-oriented *interdisciplines* have developed, such as environmental history and ecological economics (Pawson and Dovers 2003; Dovers et al. 2003).

Key differences between disciplines must be identified and reconciled for effective interdisciplinary interaction. A survey of these is not possible here, but some deserve mention. One is the spatial and temporal scales implicit in theory and method. Another concerns whether approaches to natural systems assume deterministic and linear or stochastic and non-linear processes. Some disciplines favour whole-system approaches, whereas others more naturally embrace reductionism. Assumptions about the motivations for human behaviour vary, as does awareness of the social construction of knowledge.

Two other differences deserve note. One is the degree to which a discipline is policy oriented, and thus relates to the contribution the discipline can make to policy-oriented research. Disciplines such as economics, public policy, and law are closely oriented to policy, whereas most natural sciences and some of the social sciences and humanities (e.g., psychology, history) are much less so—it is not their topic. There is a tension here. Policy-oriented disciplines may be expected to have more to say about policy. But if existing policy processes and instruments are deficient, as many concerned with sustainability believe, then the explanations of traditional economics, law, or public policy (for example) might lack purchase. The second difference is the quantitative-qualitative divide where, at the extremes, quantitative researchers cannot accept 'rigour-without-numbers' but are in turn suspected of assuming away reality to allow mathematical neatness. Away from these extremes remain problems of reconciling methods, data sources, and modes of analysis. Such deeper differences must be explored before practical issues of interdisciplinary interaction are negotiated.

We should also recognize *intra-disciplinary variation*; theory and method are not uniform within disciplines. For example, the subdisciplines of resource and environmental economics generally utilize neo-classical assumptions and methods, whereas ecological and institutional economists may not. Empirical ecologists will define problems and design analyses differently than ecosystem theorists. Black-letter lawyers approach questions differently than sociologists of law. And so on. The choice of collaborator from another discipline is a critical decision, as it will influence theoretical assumptions, problem definition, methods, data requirements, and findings.

Methodological Development and Applied Problem-Solving

Interdisciplinary research may seek to enhance understanding, but just as commonly—and often in concert with agencies or stakeholder groups—the aim will be to develop analytical methods and decision or policy-support techniques. At times, R&D may involve not only the development of techniques but also their application to actual problems. The distinction between the two may be blurred, but it is an important distinction to be aware of, as it raises questions about the role of research, the difference between research and consultancy, and responsibilities and liabilities within the policy system. The inevitably political nature of policy processes and decisions makes this more complicated than the better-understood continuum of basic applied research.

There are far too many actual or potentially integrative methods to cover here, so illustrative examples will suffice. Some methodological development involves the extension of existing approaches, such as with extended cost-benefit analysis incorporating non-market valuation or satellite physical resource accounts appended to national economic accounts. Historians and natural scientists, separately or in combination, may seek to meld documentary, oral, and scientific information to establish previous vegetation patterns or river conditions. Some methodological development may stem from a questioning of the appropriateness of merely adapting existing approaches. Examples are *multi-criteria analyses* (MCA) that do not rely on monetary valuation or integrated *green accounting* as a deeper (highly contested) integrative strategy to correct perceived deficiencies in the national accounts. All such approaches have variations and can be used in either informative or decisive modes. For example, MCA may integrate factors towards a single option or be used in a heuristic fashion to assist but not instruct decision makers. Some approaches are particular to one set of users, such as *triple bottom line accounting*, which seeks to operationalize the policy integration principle in the operations of private firms.

There is a critical link between integrative methods and deeper interdisciplinary interaction. Integrative methods can be utilized without a shared understanding of the assumptions and theoretical propositions that underpin method, and of the limits or qualifications those might entail. For example, contingent valuation, a non-market valuation technique, may be used in an integrative assessment project, with the participating scientists or managers being unaware of contests over its underlying assumptions (e.g., reliance on willingness-to-pay rather than willingness-to-avoid, Knetsch 2003). Any integration program or initiative should seek to make all proposals methodologically explicit and to encourage exposure of the assumptions that underlie methods.

Policy Processes, Organizational Structures, and Institutional Settings

The third broad means of integration is the creation of policy processes, institutional settings, and organizational structures to achieve integration of environmental, social, and economic factors. Traditional divisions across portfolios and agencies can impede integration. Again, disciplinary or interdisciplinary research may inform such design, likely through a smaller range of policy-oriented disciplines.

Many policy, organizational, and institutional remedies to fragmentation exist or have been proposed, and again only illustrative examples are given here. Approaches such as strategic environmental assessment or sustainability assessment aim to embed environmental concerns into the policy process across different sectors, significantly extending the scope and impact of the tradition of the more limited project-based environmental impact assessment. Environmental officers in non-environmental agencies (e.g., defence, water supply, etc.) serve a similar aim at a more operational management level. Placing production and conservation functions within one portfolio rather than in separate ones (e.g., a department of conservation and agriculture) is an integrative strategy that has been tried in many jurisdictions. Integrated catchment management is an integrative organizational reform. Cross-sectoral policy (e.g., oceans, biodiversity) addresses integration, as does legislation imposing responsibilities for such issues. Whole-of-government integrative strategies include environmental or sustainability offices or commissioners in some jurisdictions and environmental subcommittees of Cabinet in others. Many countries have established a multi-stakeholder national council for sustainable development to promote coordination and integration.

Communication-as-Integration

Communication represents an integrative strategy, either in and of itself or as a necessary co-element with other strategies. On the first possibility, separate disciplines, professions, and policy sectors have limited understanding of the theory, methods, and data in other domains, and straightforward communication may advance integration. On the second, communication is necessary to the success of other integrative strategies: interdisciplinary research, methodological development and implementation, participation, and policy and institutional change. All these involve new groupings of people and new flows of information and knowledge.

3.3. Integration through Participation

The modern idea of sustainability places as much emphasis on community participation as it does on environmental-social-economic integration. Indeed, the two principles are related. Like community participation, participation-as-integration requires the differentiation of ends and means. This issue is merely noted here, with the following purposes and forms of participation listed as examples:

- To integrate community perspectives into policy debate and formulation via inquiries, inclusive policy processes, deliberative research methods, representative membership of advisory committees, etc. (Munton 2002).
- To integrate community members into policy and management implementation or monitoring and evaluation via mechanisms such as community-based land management groups, honorary rangers or similar positions, co-management arrangements, etc.
- To integrate local or specific cultural knowledge with formal scientific knowledge, involving two-way flows of knowledge between community and formal knowledge systems, or collaborative (participatory) research.

There is a tendency to conflate participatory and interdisciplinary research. While they may overlap, the distinction should be maintained for clarity. An economist and ecologist may collaborate to develop an integrative method for addressing conflicts between conservation and agricultural production—a situation that is interdisciplinary but not participatory. An ecologist might undertake research on remnant vegetation in partnership with local landholders—a situation that is participatory but not interdisciplinary. Participation does not equal participatory research unless the contributed knowledge brought to the process by those participating (e.g. landholders, Indigenous owners) is treated as a valid knowledge system.

3.4. Who Integrates: Roles and Skills

Many disciplines, interdisciplinary enterprises, professions, and parts of the wider community have the ability or potential to contribute to integration in resource management. Who can contribute what to a specific process or project will vary according to the problem at hand, and this underlines the importance of the problem definition phase and of maintaining a clear problem focus in integration, however that problem is defined. To illustrate, we can consider the broad stages of a policy process and the formal disciplines of relevance. These stages are problem framing, policy framing, policy implementation, and implementation and monitoring (Dovers 2005b). In the first stage, the focus may be on negotiation of broad social goals (inviting the contribution of those knowledgeable in political science, history, demography, philosophy, etc.) or on environmental change (inviting the contribution of those in the fields of ecology, atmospheric chemistry, hydrology, etc.). In policy framing and implementation, disciplines such as public administration and law have particular roles. In the final stage, where policy is evaluated and where policy monitoring and environmental monitoring must be linked, there is a role for a wide range of disciplines in the natural and social sciences. Taking a complete view of the policy process forces recognition of the need for at least multidisciplinary and probably interdisciplinary R&D. As well as formal disciplines, a wide array of interests and other knowledge systems throughout the community are relevant throughout the policy process.

A crucial factor in problem definition in the research process is that often a single individual or limited group defines the problem and then seeks support and involvement. If the contributors to the research project together lack an exceptional grasp of many disciplines, the problem definition and thus inputs to the R&D process may be constrained.

3.5. Scale and Integration

To understand and manage linked environmental, social, and economic systems, those who practise integrated research must deal with interactions across multiple spatial and temporal scales. Some elements of the task are well appreciated, such as disjunctions between political and ecological or hydrological boundaries or between temporal scales over which ecological, political, and economic processes operate. Such disjunctions represent research as well as practical policy and management challenges.

In interdisciplinary research, this issue can be subtle, as scales are embedded in the theory and method of different disciplines. Like much about disciplines, this may involve assumptions and determine method and data in ways not apparent to those outside a discipline and even taken for granted by those within. For example, the spatial scale of the law is defined by jurisdiction—for anthropology, it may be culturally or ethnically determined; for hydrology, by watersheds; and for economics, by individuals, firms, national economies, and trading systems. The temporal scale of the law is determined by the enactment of statute law or by common law precedent—for ecologists in a range of ways and differently again for historians. As with other disciplinary features, the task is to seek clarity in the stages of problem definition and research design through justification of the chosen scale with reference to the problem at hand and to problems inherent in transferring information or findings across scales.

3.6. Interconnections and Learning

Categorizations such as the preceding recognize multiple meanings of the term *integration*, which is often used as if it has a single meaning. A subtle understanding of the term is essential to efficient and effective research and policy. However, the blurred boundaries and interconnections between the dimensions of integration are as important as what separates them. For example, different purposes may be interrelated. Integrated research that seeks to deepen understanding of linked phenomena can inform policy design to manage such phenomena, if this potential has been incorporated into the research approach. Similarly, theoretical inquiry may be a necessary precursor to methodological development.

The benefits of recognizing interconnections include the potential efficiency of achieving multiple objectives through single investments and a reduced likelihood of poor outcomes through incorrect problem framing. It is easy but insufficient to claim the general relevance of integrative research to resource management. It is necessary to specify precisely how the connection can be achieved, whether through research design or communication.

Given the differing purposes and forms of integration, many individuals and organizations will engage in a fragmented experiment over many years. It can be expected that these participants will have difficulty maintaining coherent directions and a shared awareness of theoretical, methodological, and practical developments. This suggests that coordination and communication will be important.

4. PRACTICALITIES AND PEOPLE: MANAGING INTEGRATED RESEARCH

Unlike the first three sections, which have been general in tone, using hypothetical or typical examples aimed at clarifying key terms, concepts, and their implications, this section addresses the *realities* of managing integrated research when it involves multiple interests and forms of integration. The section uses a case study of a long-running and successful R&D program, supplemented with personal insights into the reality of integrative research in a human landscape.

Integration in theory is complex, and in practice it is frustrating, time-consuming, costly, and publicly degrading. This may explain why integration is also exciting, rewarding, and ultimately worthwhile. The experience of one of Australia's environmental research funding bodies, Land & Water Australia, demonstrates that integration is both a necessary evil and a kindly matron. Ultimately, the aspect that prevails comes down to the managing of relationships, at both the institutional and personal levels.

In 1992 managers in Land & Water Australia perceived the need to address dryland salinity across different jurisdictions, disciplines, and geographic localities in a single, integrated research program. The National Dryland Salinity Program (NDSP) was created, with a wide range of federal, state, industry, and research partners.[3] Analysis of previous failures to deal adequately with this growing national threat showed that the essential knowledge-generation efforts were institutionally and geographically disparate and largely competed with one another for limited research funds. Remote-sensing techniques would be researched in one catchment, for example, landscape modelling in another, agronomic solutions in yet another, while economic and social research would hardly be looked at.

With the NDSP was born the idea of using focus catchments to provide foci for scientific integration while at the same time enabling local stakeholders to participate in the management and conduct of research. Five catchments across Australia were selected (on grounds that were more about politics and equity than science). The lessons learned from this process included the following:

Problem definition. Everyone has a different perspective on what the problem is, and in many cases, those who 'see the big picture' see a very different big picture. For example, in a NDSP workshop, a bureaucrat painted his big picture based on how catchment management fitted into a broad, macro-policy setting. A catchment manager showed how policy was only one perspective within a large and diverse view of community relationships. A researcher, newly acquainted with emerging tools to integrate economic, social, and biophysical data, argued that policies, relationships, and 'hard facts' could indeed be quantified and structured within complex systems models, from which good policies and practices would flow. And finally, a farmer simply said, 'This is the big picture: I've got to eat, got to pay bills, got to educate the kids, got to replace the machinery, got to second guess God about the weather and got to look after not just the farm, but the catchment and the rest of the country too!' Bringing different perspectives to bear on a problem and recognizing the differences in perspectives are critical.

Collaboration. Getting groups to agree on a definition of a problem is difficult, but getting agreement to collaborate is even harder. For many NDSP partners, collaboration meant having to find and allocate resources, convince hierarchies about the importance of a partnership, and agree that not every agency can be the 'lead' agency. While consensus was reached about a common NDSP goal, it had to be recognized that this goal was not always fully consistent with the goals of the individual collaborators. 'What's in it for us?' was a more common thought than 'How can we contribute?' A legal framework establishing a framework for the collaboration, including specifying management and intellectual property arrangements, led to less interference by institutional processes in the research processes. Without collaboration at the management level, it is hard to create an environment for integration at the research level.

Integration frameworks. The NDSP tried several frameworks for integrating different research knowledge over 10 years. Modelling proved an expensive challenge. Do we build a new model or refine or drastically alter existing ones? Should the model be one-dimensional or two? One based on understanding processes, or one providing management options? It seemed that every researcher was married to a particular model. All variations were supported. In the eleventh and final year of the program, the most effective model for integration was found—three separate manuals for three different audiences. The manuals provided a rallying point, and the process of developing them relied on teamwork and synthesis. Debates had timelines governed by the publication date. The beauty of the manuals was that they posed questions for researchers to answer, but in such a way that no one discipline could answer them satisfactorily. In this case, the successful strategy to forcing integration was to define the task from the demand perspective rather than from the perspective of multiple contributing disciplines.

Resources. Interdisciplinary research is costly, especially if modelling requires field-based data collection and validation. Taking into account social, environmental, and economic factors required investment in research not traditionally supported in, but critical to, salinity management. Having to attend additional meetings and workshops and working on 'other people's territory' (especially when this was far from the home base) meant enormous strains on time and budgets, but proved an indispensable means of getting integration to occur face-to-face. Interaction, communication, and coordination are investments in integration, not costs.

Pooling. Pooling resources is more than a gesture of goodwill. The funding model for the NDSP evolved over time but was most effective when the different organizations were able to pool their resources in a common funding account. This included pooling management and technical expertise to ensure accountability for the expenditure and allocation of pooled funds and the rigour of the science undertaken. Placing one's resources into the trust of a collective is a starting point for breaking down barriers between institutions and disciplines.

Conflicting masters. Researchers involved in integrative research are inevitably placed in an arena of conflict. The NDSP supported over 50 projects, each of which was associated not only with the NDSP, but also with one or more programs of the partner organizations. For example, a catchment characterization project supported by the NDSP was also supported under a partner's basin strategy program, another partner's state salinity program, and another partner's hydrology program. Where activities were embedded in ongoing programs of collaborating partners, these activities were conducted within an environment of conflicting infrastructures, agendas, and hierarchical allegiances. This required support by respective management interests to enhance researchers' capacity to navigate complex institutional boundaries. In many cases, however, researchers could see what needed doing and with whom research was best shared, and they just got on with it, sometimes completely under the radar screen of management.

Information management. The NDSP included a technical committee that helped to ensure that common approaches and protocols were used across projects to assist the integration of information. This committee was largely free of the responsibility of conducting research for the program, but had a very close association with it. Because it had the role of directing what research was undertaken and how, it was able to influence the degree to which integration took place. For example, the committee directed that a catchment characterization system, which was developed to map regional groundwater flows, be accepted as the basis for underpinning projects as diverse as engineering, economic cost analyses, saltland production, and local government policy. A well-respected guiding hand can play a significant role in facilitating integration across institutions, disciplines, and regions.

Time frames. Integrated research will always take longer than planned. In the case of one highly complex catchment modelling project, which had a life of five years granted to it on the basis that the establishment might take time, getting agreement to the conceptual framework took over two years. This time was needed to develop team ownership, shared understanding, team trust, and something resembling a common language. The project did not finish on time, and was better for it. Investing time to build relationships needs to be a legitimate part of the research process.

Communication. Here is where successful integration starts and ends. The most powerful tool of the NDSP was its communication network. This operated at several levels: among management, among technical and communication specialists, among stakeholders, and, most importantly, among all of these groups. One-quarter of the NDSP's budget was dedicated to communication. People learned from sharing experiences. Communication materials were developed with different users in mind. Researchers liked to share information with other researchers; they didn't mind at all knowing what was going on in other disciplines that were dealing with the same substantive issue. Farmers liked to share their success stories with other farmers, and they also liked to get a comment or two from researchers about why something they were doing actually worked. Communication was facilitated through print, websites, list-servers, workshops, and conferences. Letting people ask questions and demand answers stimulated more integration than any model could have.

The case of the NDSP indicates the realities of practice that exist beneath the different kinds of integration defined earlier in the chapter. The following personal reflection, by program manager Richard Price, is a more direct, plain-language insight into what happens under an initiative such as the NDSP. It is important to remember that researchers and stakeholders are human beings and that they behave as individuals, not as either theory or method instructs or expects them to behave.

> I will never forget my five-year association with the Liverpool Plains of northern New South Wales. One moment it was a backwater of the state, with the biggest research problem being 'To till, or not to till?' Suddenly salinity was the big issue, with the major problem being soil compaction from every researcher and his dog wanting to tread all over the farm paddocks in the district. In one foul swoop, the Liverpool Plains had become a research focus catchment. Participatory research, interdisciplinary research, and multi-organizational research were the order of the day! 'Here is where the money is at!' said the funding bodies. 'Then that's where we will go,' said the research agencies.
>
> The major project had a title like 'Integrated catchment modelling to support community decisions in the Liverpool Plains.' Eight major research organizations, including four federal, two state, and two university groups, were involved. All up, 35 researchers covering fields such as soil science, hydrology, economics, systems modelling, plant physiology, geology, remote sensing and the like were involved. One organization had been given the contract on behalf of a research consortium and was expected to coordinate all the others.
>
> At first, no-one objected to coming on board of such a complex project. For some organizations, collaborative projects meant attracting further resources into the organization. Indeed, for these organizations it was the only way to attract resources—collaborate or perish! Salinity is a complex issue, and it seemed just common sense to work across disciplines in the one region.
>
> I wondered, however, if we were overcomplicating something that perhaps had no possible solution—that if we threw enough disciplines at the problem it

would just go away. Still, it was good form to collaborate, and meant that you were open to new ideas. This project offered to assemble some of the best minds in the salinity game, and bring their combined wisdom to bear on what we thought was a relatively simple catchment.

However, dealing with so many institutions and disciplines had its downside. Arguments proliferated about resource distribution, which models to use, who should report to who, and which hierarchy to follow. We also had to take care to recognize whose territory we were on at any point in time. Egos ran rampant. Voices were raised. Tempers flared. Researchers were at it hammer and tongs. This was science at its best! And things would have run well enough too, except that this was also supposed to be a participatory research project. The debates we had, the fine detail we fought over and the language we struggled with all took place in community halls and on farmers' paddocks. No wonder the community was frustrated, looking on in astonishment at the intrinsic messiness of interdisciplinary research. What was normal for us looked totally unruly to them. Indeed, it was unacceptable and they told us as much.

But on reflection, it wasn't normal for us. We were struggling with the notion of interdisciplinary science at the same time we were struggling with the notion of participatory research. The complexity of integration had increased exponentially. I realized there were so many of us, having so many arguments, because we were responding to a community with so many different perspectives. We learnt quickly that there is no one community, with only one perspective, and that realization ultimately became the major research challenge. And above all, we learnt quickly that we are all amateurs when it comes to integrated research.

The project progressed over the years through paths rocky, turns unexpected, and pit-stops unplanned. It took over two years before the research team was relatively comfortable with one another. Suspicions about each other's motives had dissipated. The community was posing questions that meant something to all of us, even though it meant something different to each. We had a basis to work from, and by the time we had agreed upon a conceptual framework, it came as a hard blow to be told that we had finally reached the starting blocks and not the end point. Still, a sense of camaraderie had developed from our shared trials and tribulations, and the remainder of the project ran smoothly.

It became apparent over the years that we were beginning to work in a very crowded territory. What was novel at the beginning of the project was becoming more commonplace by the end. The managerialists of the world had taken control of the boardrooms, and every organization was expected to deliver not just outputs, such as new technologies, but also deliver outcomes—evidence that clients were using the technologies. This was great in theory, but it meant that every organization was seeking changes in client's practice, and so every institution started to develop its own communication capacity, each trying to engage the same clients. Also, everyone wanted to take the lead in playing the role of coordinator, so they could be seen to be the one driving the outcomes. It didn't seem to matter if the research group was a university or a government regulatory body with a research arm. More and more resources of different agencies seemed to be going into dealing with the same stakeholder, and less seemed to be going into research.

I was wondering if we had lost sight of vertical integration between research institutions in wanderlust of horizontal integration with the community. Ironically, are we becoming less efficient in the use of limited research funds, and as a result providing less value to the very community we serve?

This reflection portrays the sort of reality often missing from articles advocating integrative research for resource management or reporting positively on only the findings of an integrative program. We now turn to a set of principles for integrative and/or interdisciplinary research, which, although well based on theory and evidence, also conveys some of that difficult reality.

5. CONCLUSION: GUIDING PRINCIPLES FOR INTEGRATED RESEARCH

There is no single path to integration, because there are multiple purposes and forms of integration. Design of integrated research needs to be guided by principles that reflect this complexity. The following extends the 'essential elements of interdisciplinarity' proposed by Barnett et al. (2003) as a set of principles that brings together the discussions in this chapter:

- A *problem focus*, whether applied, theoretical, or methodological. Without a clearly defined problem, integrated research can only succeed through luck. Explicit problem definition encourages early consideration of the skills and perspectives required.
- A wariness of *the dangers of capture* by singular or partial policy objectives. Sustainability is a long-term problem that is pervaded by uncertainty and deals with non-linear systems, and integrative research should not be driven solely by immediate agency agendas.
- A *critical, reflexive capacity*, including recognition of the normative content of claims to knowledge. Given the magnitude of the task and the uncertainties inherent in it, sharp and constant evaluation of integrative initiatives is required. However, so as not to descend into a non-constructive relativism, this critical stance needs to be informed by ecological, economic, and political realism.
- *Openness to other disciplines*, theory, methods, and arenas of inquiry, as well as to cognate policy sectors and knowledge systems other than formal disciplines (lay, professional, Indigenous, etc.).
- A *systems orientation*, appreciating the whole rather than only parts and encompassing both quantitative and qualitative constructions of systems. Essential to this is appreciation of key systems properties such as feedbacks, path dependency, thresholds, and time lags.

- A close appreciation of *multiple and dynamic spatial and temporal scales*, including a capacity to account for historical determinants of modern situations.
- Appreciation of the *personal/group qualities* required for interdisciplinary work, as well as of the balance of risks and rewards in crossing disciplinary and other boundaries. Previous patterns of interaction, incentive, or reward are unlikely to be suitable, whether in research institutions, policy agencies, community groups, or private firms.
- The need to *recognize multiple purposes of integration* (understanding linked phenomena, informing policy and management, implementing policy and management, participation) and potential interconnections and synergies between integrative projects driven by different purposes.
- Recognition of the close connection between *problem definition and the varying contributions of different disciplines* and other knowledge systems in order to apply specific mixes of skills and understanding to specific problems.
- Recognition of *intra-disciplinary variation*, given significant differences in implicit scale, problem definition, theory, method, and data requirements within as well as between disciplines.
- Recognition of *communication as central to integration* in terms of communicating new integrative outcomes to potential users and of encouraging integration through communication of specialized perspectives to new audiences.

These principles are generic but operational enough for translation to specific contexts. They may render integration more problematic, which is consistent with the reality that integrative research, policy, and management are significantly more difficult and complex than non-integrated approaches, demanding that we undertake more and do it better, as well as differently.

NOTES

1. The paper draws on and extends Dovers 2005a.
2. The humanities have not been included here, although arguably they should be (e.g., the role of historical analyses of landscape conditions, human motivations, and policy experiences; see Dovers 2000).
3. Richard Price managed the NDSP for 11 years. For a synthesis of the activities and products of the program, see Robins 2004, and for an analysis of another integrative program see Price 2003. This account draws on this close personal involvement.

REFERENCES

Barnett, J., H. Ellemor, and S. Dovers. 2004. Interdisciplinarity and sustainability. In S. Dovers, D. Stern, and M. Young (eds), *New dimensions in ecological economics: Integrative approaches to people and nature*. Cheltenham: Edward Elgar.

Common, M. 2003. Economics. In E. Page and J. Proops (eds), *Environmental thought*. Cheltenham: Edward Elgar.

Connor, R., and S. Dovers. 2004. *Institutional change for sustainable development*. Cheltenham: Edward Elgar.

Dovers, S. 1997. Sustainability: Demands on policy. *Journal of Public Policy* 16:303–18.

——— (ed.). 2000. On the contribution of environmental history to current debate and policy. *Environment and History* 6:131–50.

——— 2005a. Clarifying the imperative of integration research for sustainable environmental management. *Journal of Research Practice* 1 (2): article M1. http://jrp.caap.org/content/v1.2/dovers.html.

——— 2005b. *Environment and sustainability: Creation, implementation, evaluation*. Sydney: Federation Press.

Dovers, S., D. Stern, and M. Young (eds). 2003. *New dimensions in ecological economics: Integrative approaches to people and nature*. Cheltenham: Edward Elgar.

Dryzek. J.S. 1987. *Rational ecology: Environment and political economy*. Oxford: Basil Blackwell.

Knetsch, J. 2003. Environmental, ecological and behavioural economics. In S. Dovers, D. Stern, and M. Young (eds), *New dimensions in ecological economics: Integrative approaches to people and nature*. Cheltenham: Edward Elgar.

Munton, R. 2002. Deliberative democracy and environmental decision-making. In F. Berkhout, M. Leach, and I. Scoones (eds), *Negotiating environmental change: New perspectives from the social sciences*. Cheltenham: Edward Elgar.

Pawson, E., and S. Dovers. 2003. Environmental history and the challenges of interdisciplinarity: An antipodean perspective. *Environment and History* 9:53–75.

Price, R.J. 2003. Identifying social spaces in the Sustainable Grazing Systems Program. *Australian Journal of Experimental Agriculture* 43:1041–59.

Robins, L. 2004. *Dryland salinity and catchment management: A resource directory and action manual for catchment managers*. Canberra: Land & Water Australia.

Schoenberger, E. 2001. Interdisciplinarity and social power. *Progress in Human Geography* 25:365–82.

UN (United Nations). 1992. *Agenda 21: The United Nation's Programme of Action from Rio*. New York: United Nations.

CHAPTER 4

Governance for Integrated Resource Management

Ann Dale and Lenore Newman

> 'Panarchy', an odd name, but one that is meant to capture the way living systems both persist and yet innovate. It shows how fast and slow, small and big events and processes can transform ecosystems and organisms through evolution, or can transform humans and their societies through learning, or the chance for learning. The central question is what allows rare transformation, not simply change.
>
> C.S. Holling, 2003
> Starker Lecture Series, Oregon State University

INTRODUCTION

Over the past 20 years, there have been increasing demands in Canada for greater public participation in decisions normally regarded as the legitimate prerogative of the state. It is not clear whether these demands stem from dissatisfaction with the way decisions are made by the current institutions of governance, whether they reflect a belief that elected officials lack the ability to represent concerns, or whether they are the result of profound cultural and social change. Regardless of cause, many Canadians are questioning the ability of their governments and its officials to represent the diversity and complexity of an increasingly plural society (Alabo, et al. 1993) and are demanding more control of and participation in political decision making.

This shift in public values goes beyond demographic groups traditionally associated with the political process. There is a demand for the integration of voices that previously have not been heard, such as women, immigrant and visible minorities, the GLBT (gay, lesbian, bisexual, and transgendered) community, and the disabled. The values of post-industrial society are fundamentally different from those of previous generations; they are much more diverse and plural (Pross 1992; Hoberg 1992; Versteeg 1992). Inglehart (1990) has argued that two major components of these values, based on Maslow's fixed hierarchy of individual needs and the need for self-actualization, are manifest through, first, the high value to having more say in determining the course of public policy and, second, having a secure right to express one's views about issues of public policy without fear of punishment.

This increased interest in a greater say in the political process is not taking place along traditional community lines. Several researchers have argued that sources of social capital such as community groups are failing to thrive as memberships shrink (Putnam 2000; Jacobs 2004) and voter turnout continues to trend downward even in very high stakes elections. However, though we may be, as Putnam says, 'bowling alone', we are engaging in a different sort of community building. Most Canadians are members of multiple and overlapping communities that include not only communities of place, but also communities of practice (Lesser and Prusak 2000). Such communities include those based upon professional affiliation, shared interests, and networks (CPRN 2003). We choose our communities in post-industrial society, and if the town hall is emptying, the electronic agora is filling. These chosen communities provide an outlet that traditional political channels have yet to provide; post-industrial citizens don't want representation, they want agency in integrated decision-making processes. Of particular interest to resource management is the ability of these communities of choice to transcend geographical boundaries; the 'global citizen' is concerned over the treatment of resources in remote corners of the world and has far greater stakes in how these resources are used and managed.

This chapter explores the intersection of the increasing demands from diverse communities for input into political decisions that affect them; it also looks at the need for evolving, multi-scaled approaches to sustainable development, specifically with reference to resource management. It is a fortunate piece of self-organization that a civic desire to increase local involvement in issues such as resource management is arising at a point at which strongly compartmentalized control is failing to meet the challenge of the complex, turbulent environments (Emery and Trist 1972) facing all human societies. Unless governments at all levels move towards integrated decision-making processes, enlarged policy-making processes, and more inclusive models of governance that involve the greater plurality, sustainable development will remain an elusive 'grail'.

AGENCY AND PUBLIC PARTICIPATION

There are several definitions of *agency* in use. Harvey (2002) defines agency as 'the capacity of persons to transform existing states of affairs.' Agency has also been defined as 'the ability to respond to events outside of one's immediate sphere of influence to produce a desired effect' (Bhaskar 1994). For social capital to mobilize, there must be an invocation of agency (Newman and Dale, forthcoming), and citizens are less inclined to hand that agency solely to elected officials. Further fuelling these demands for changes in political decision making is the sophistication of post-industrial society. The interested publics no longer readily accept decisions made by their political leaders on an ad hoc basis; rather they demand explicit criteria and explanations about their leaders' approach and logic. Both technological development and an information explosion have influenced the culture of post-industrial society. The rapidity and degree of change in modern society is unparalleled in humankind's history.

Encouraging community agency in the area of sustainable resource management requires cooperation on a number of levels. There are some limits to agency creation: technological limitations inhibit possible actions; differences in feedback scales obscure our ability to measure the effects of our actions; and our options might be sanctioned by more powerful actors (Dietz and Burns 1992). All of these limitations can be mitigated to some degree through government facilitation. All levels of government can act as distributors of appropriate technology, as monitors of key feedback processes, and as partners in the integrated management of resources.

Coupled with the shift in community values about greater public participation is another significant feature of post-industrial society: since the 1970s, there has been growing concern and expanded knowledge about environmental threats and the corresponding damage to the biosphere (Caldwell 1974). Two major international milestones heightened this awareness, the 1972 Stockholm Conference and the 1992 United Nations Conference on the Environment and Development (UNCED). Wedged between these two international events was the landmark 1987 Brundtland Commission report that signalled the urgent need for the world to integrate environmental considerations into economic decision making if humanity was going to survive beyond the twenty-first century. The commission argued that the ability to choose political paths that are sustainable requires that the ecological dimensions of policy be considered at the same time—and on the same agendas and in the same national and international institutions—as the economic, trade, energy, agriculture, industrial, and other dimensions (Brundtland 1987). Even earlier, the Macdonald Commission (1985) warned that '[i]t will be essential in the decades ahead to integrate environmental decisions and economic decisions, for there is, in the Commissioners' view, no ultimate conflict between economic development and the preservation and enhancement of a healthy environment and a sustainable resource base.'

Further complicating the holy grail of increased public participation and transparency in decision making is the growing imperative to integrate environmental and economic decision making, and the consequent push to move from the traditional sectoral approach to multidisciplinary approaches (Bregha et al. 1990; Caldwell 1974). Two additional pressures stimulate this demand: first, current public institutions seem unable to respond quickly to the emerging environmental imperatives, largely owing to their solitudes, silos, and stovepipes (Dale 2001), and second, there is a growing convergence and acceptance of sustainable development and, in particular, integrated resource management as a governance approach. Governance for integrated resource management demands a holistic systems approach and understanding of the relationships between natural and human systems (Shackley and Wynne 1995; Hill 2005). More importantly, it demands integrated decision making, but this integration calls for a more complex and reflexive form of analysis (Rothman and Robinson 1996) and policy development capacity.

GOVERNMENT RESPONSES

How well have our institutions, and in particular our governments, responded to our calls for empowerment, cooperation, equity, sustainable development, and security (UNDP 1997)? How have they responded to our requests for greater transparency and accountability (McFarland 2003) and for greater internal coherence through policy integration, involvement of stakeholders, improved knowledge management, and improved policy coherence and integration for sustainable development (OECD 2002)? One could assume that given the declining participation in voting, especially among youth, 24 per cent in Canada and 18 per cent in the United States, the answer is clearly not ideal: we are indeed 'bowling alone'. If people do not feel that their voice is making a difference, they become disempowered and then disengage (Dale 2005). Thus, a key feature of social capital for institutions might lie in their capacity to restore agency and trust to citizens in their relations with formal structures and processes (Dobell 1995).

How, then, must government engage the multiple and contending outside stakeholders if, indeed, it is to participate in enlarged decision-making contexts? Such contexts are unlikely to be established unless a new socio-governmental context emerges through the spread of trans-bureaucratic organizations and the creation of a common ground around the necessary changes (Emery and Trist 1972). We need to move from closed to open policy-making processes, from issues that are single sector and domestic to ones that are transdisciplinary and global. We need to move from government as controller and monitor to government as catalyst and leader, from citizen participation based on exclusive invitation and exclusion to citizen participation based on rights, competency, and responsibilities of inclusion. This will involve several role shifts: policy analysts must evolve from being technical specialists to being just individual members of transdisciplinary teams, and management must shift from models that are primarily vertical to clustered structures capable of responding to negative feedback. In order to achieve working heterarchies of control, other shifts must take place. These heterarchies necessitate a move from a homogenized perspective to a diversity of values, from a horizon that is short-term and reactive to one that is long-term, proactive, and multiple in time, place, and scale, and from adaptive, reactive management to proactive, responsible, dynamic management.

Unfortunately, the answer to how well government has responded to these new imperatives, and in particular to integrated response management, is that it has not responded well. Canadian governments are still organized around problems that have changed fundamentally since the creation of the British North America Act in 1867. Thus, silos such as departments of agriculture; energy, mines, and resources; environment; and finance are inadequate to respond to integrated, interdependent global issues such as climate change, biodiversity conservation, and integrated resource management that span departments, cross sectors, and demand transdisciplinary thinking and solutions. Moreover, the vertical silos created by this traditional departmental structuring lead to fiefdoms that are antithetical to the coherent policy design for integrated decision making so critical to the implementation of sustainable development. We need to move from government structures that are problem based to ones that are issue based and capable of dynamically self-organizing around domains of appreciation.

One such domain of appreciation is sustainable development. Sustainable development can be regarded as a process of reconciliation of three imperatives: (a) the ecological imperative to live within the global biophysical carrying capacity and to maintain biodiversity; (b) the social imperative to ensure the development of democratic systems of governance to effectively propagate and sustain the values that people wish to live by; and (c) the economic imperative to ensure that basic needs are met worldwide (Dale 2001; Robinson and Tinker 1997). Sustainable development requires restructuring of government structures, cultures, processes, and tools, as it is becoming increasingly obvious that the tasks of postmodern governments now extend from regulating the present to creating the enabling conditions for the future (Emery and Trist 1972). Thus, it is clear that institutions structured within stovepipes, such as disciplines in universities and divisions within departments, coupled with silos between departments that are dominated by problems generated by the dominance of departments of finance over departments of environment, are antithetical to integrated decision making for the reconciliation of these three imperatives.

A STRUCTURAL PROBLEM

The result of all of the above is that despite considerable expertise, science, planning exercises, program reviews, and task forces on what is needed for change, we have a serious implementation gap in our ability to respond to the critical issues facing contemporary society, and we are far from integrating environmental imperatives and the diversity of voices waiting to be heard. One of the major reasons for this implementation gap is the gridlock in the planning and implementation processes for decision making. This gridlock is not due to a lack of research, knowledge, or information in communities, but rather has arisen as a result of the solitudes, silos, and stovepipes (Dale 2001) that characterize the research, business, and governance sectors. The problem is multi-faceted and involves, among other things, a lack of coherent dialogue, a lack of congruence between political levels, little political will, and the lack of a sustainable development ethos among government levels and community stakeholders. Others have referred to fundamental disconnections—between federal, regional, and local governments, between rural and urban communities, and, critically, between the business and research communities (Bradford 2003; Dale 2001).

In order to build the needed partnerships among stakeholders, we must ensure that information flows between the solitudes discussed above. In order to take knowledge from a university setting to a community setting or from the government sector to local stakeholders, we must engage in transdisciplinary research and, ultimately, transdisciplinary decision making. Formally defined, transdisciplinarity bridges scholarly enquiry and the sphere of tacit and experiential knowledge (Horlick-Jones and Sime 2004). Further, it is based on the assumption that practices must evolve to match the complexity of the issues facing today's scientific community, and it aims to narrow the gap between research and decision making (Ramadier 2004).

Thus, it is not sufficient merely to improve current modes of operation (Gunderson and Holling 2002). Fundamental transformations are required in all sectors of society; these include changes to our social, economic, and governance structures, coupled with shifts in cultures and practices. We need institutions capable of self-organizing around issues, rather than institutions that prefer to focus on outdated problems and conflicts that no longer reflect modern reality. One approach is adaptive management, but it is our contention that many of the issues we face require the ability both to adapt (in reality, human and natural systems are now co-evolving, given the great impact of human activities on the biosphere) and to redesign, proactively, our systems and redirect our impacts—an approach we refer to as 'responsive management'. What is the difference between the two? Responsive management is the ability to proactively and strategically determine major trends and issues in advance of experiencing their actuality and to deliberatively design processes and policies in order to respond and redirect human activities at the micro, meso, and macro levels. What would institutions designed for integrated resource management look like?

INSTITUTIONAL CHARACTERISTICS THAT SUPPORT SUSTAINABLE DEVELOPMENT

Sustainable development is not window dressing that can be 'tacked on' to a resource management institution. It is a complex paradigm that must, if it is to be realized, be deeply embedded in all facets of an institution. Sustainable development principles must first be articulated and integrated into all aspects of the institution's approach to resource management. Each part of an institutional system should interpret its mandate broadly to take into account all three dimensions of sustainability—the ecological, the social, and the economic. Each part of an institutional system must work to identify the plurality and diversity of the values, theories, and practices associated with the resources the system addresses and/or the services it delivers. The institutional system might, for example, employ the principle of 'full cost accounting' in assessing the outcomes and impacts of decisions, or it might examine the values inherent in optimization versus maximum yield strategies. In embracing the uncertainties and complexities inherent in sustainable resource management, each part of an institutional system engages a precautionary principle: each part recognizes that ecological, social, and economic limits exist, though they may not be definable, and thus takes a cautious approach to solving problems and making decisions to ensure that outcomes are within those limits.

To realize integrated resource management, all decision making fundamentally should integrate ecological, social, and economic imperatives within a guiding framework articulated for each resource an institution is attempting to manage. Social imperatives, though often less quantitative, cannot be ignored. Institutions should ensure that the costs and benefits of the decision-making processes and their outcomes are distributed fairly among those affected; they should also provide feedback mechanisms for those who feel that their interests have been overlooked. This feedback should be evaluated in future policy deliberations.

One of the challenges institutions face in pursuing a sustainable approach to resource management is the need to develop the ability to contextualize feedback; the system should have the capacity to respond equally and the analytical ability to reconcile competing ecological, social, and economic feedbacks. Of particular importance is the ability to recognize and respond in a timely fashion to negative feedbacks from ecological systems, particularly with respect to the loss of diversity at multiple scales and levels. This involves 'a strong focus on interactive effects, second and higher order consequences, non-linearity, thresholds, changes in state, diversity and resilience' (Rothman and Robinson 1997, 6). It also demands enlarged policy development contexts, as well as proactive rather than reactive policy development and the critical ability for strategic long-term multiple scenario development.

Institutions also have to be able to recognize and respond to the differing time, place, and scale phenomena of natural systems and human activity systems, recognizing that the biosphere imposes absolute limits on human activities. Accordingly, decision-making systems should reconcile an ecological framework of spatial boundaries with socio-political boundaries, taking into account the finite limits on place and scale imposed by the biosphere. A cautious approach is to live below, rather than near or at, those limits in order to optimize the resilience of both the natural and human systems, maintaining a dynamic balance instead of being subject to the boom-and-bust cycles of resource exploitation that have characterized our history.

In terms of time, Holling et al. (2002) refer to the panarchy model's ability to respond to both fast and slow variables. The fast variables are economic, and the slow variables are educational and cultural. One is left with two questions: how do we recognize and communicate the importance of investment in the slower variables, and how do we combine the advantages of encouraging fast variables without threatening the slow variables (Carpenter et al. 2001). In terms of integrated resource management, this means that decision making must be able to hold a creative tension between short-term exploitation and longer-term maintenance of the resource for future generations.

It is worth noting here that the International Panel on Climate Change (Mortsch and Mills 1996, 34) identified five critical features for integrated assessment: (1) formal integrating models; (2) expert panels; (3) formal models of subcomponents linked through external, judgmental combinations of results rather than through formal integrating models; (4) collaborative, interdisciplinary research teams interacting to develop knowledge-sharing skills; and (5) simulation/policy exercises using existing information to develop hypothetical contexts for discussion between researchers and policy-makers.

A NEW MODEL FOR GOVERNANCE

Sustainable development issues, therefore, necessitate a different style of stakeholder engagement, one that will change how an institution approaches a resource management issue. One can often tell from its interaction with stakeholders whether an institution has absorbed all three aspects of sustainable development. Once versed in social sustainability, an institution will engage in decision making that involves a plurality of the interests concerned. Diversity of representation in processes, decisions, and actions, plus a multiplicity of approaches, needs to be dynamically employed. This cannot just be a superficial attempt at transdisciplinarity, engaging only 'known entities'. Each part of the institution should provide opportunities for a diverse majority of affected interests to participate in resource decision making and to undertake actions that affect their future, while sharing responsibility for the future viability of the resource in question.

This transdisciplinarity should not be something that is recreated for each concurrent process; it should be intrinsic. Each community engagement process should gather together a multiplicity of stakeholders who can bring relevant experiences to bear on the specific issue, in particular natural and social scientists, public policy practitioners, the non-governmental community, and the users and takers of the resource. The institution as a whole should practise holistic embeddedness—that is, it should recognize that it is a small part of a larger grouping, that it is embedded in larger interacting systems rather than being the dominant system. This recognition provides a foundation for valuing history, intergenerational and global equity, and the needs of others (Josselson 1996). Each part of the institutional system is a part of a larger whole that furthers democracy by strengthening civil society, for it is clear that a dynamically engaged and literate citizenry is crucial for more sustainable futures.

The way an institution treats information reveals much about the degree to which it accepts a sustainable development paradigm. The integrity of information, particularly the negative feedbacks from ecological systems, is critical to the responsiveness of resource management institutions. Unbiased and detailed timely information is important, and knowledge of the limitations of information is equally so. Institutional systems must recognize the complexity of, incomplete knowledge about, and uncertainty inherent in living systems. They must also recognize that no one sector can solve the complex societal issues involved. Rather, their role is to stimulate networks of collaboration around 'domains of interest' to find solutions and undertake concrete actions.

A responsive institution does not only encourage and practise sustainable resource management, it also self-manages in a sustainable way. This process can take many forms, and all of them reflect aspects of a deeply embedded philosophy of sustainable development. For example, components of the institution should work together harmoniously to avoid implementation gaps and policy incoherence. Each part of the institutional system must recognize linkages with other parts of the system, must seek to harmonize its activities with those of others, and must promote a coordinated approach to achieving overlapping goals. Part of this process should involve a search for institutional efficiencies that would allow the institutional system to reduce overlaps and redundancies in the mandates and activities of its component parts; individual parts of the system should not duplicate the efforts of other parts. This criterion recognizes, however, that some degree of overlap is necessary to support integration and ensure that the system has the robustness needed to respond to unexpected events.

Internal resource allocation is very important. Each part of the institutional system requires a sufficient mandate and level of staff and resources to run processes, make decisions, implement results, and monitor or review outcomes as necessary to achieve its objectives. The operation of the system can then produce meaningful results from the perspective of both those operating the system and those who receive the services provided by the system. When each institutional component has a sustainable workload, it can respond in a timely fashion to the constituency it serves, and can provide mechanisms by which that constituency can directly hold individuals or groups responsible for a decision or action.

Perhaps institutions can most effectively blend adaptive and proactive approaches through their own internal processes. Institutions need to be perceptive and proactive, looking for present and future opportunities and challenges. They must prioritize their actions on the basis of an assessment of the scope of impacts, the irreversibility of decisions or actions, and the degree of urgency; in addition, they need the capacity to address short-term crises, undertake long-term planning, and also anticipate and respond to issues that occur at 'in-between' speeds.

Each institutional system requires the capacity to keep up with changing values and knowledge and to evolve decision-making processes in response. Each system requires the mandate and tools needed for self-reflection, self-evaluation, and self-modification, and each must show leadership in questioning not only the way things are done, but also whether the 'right' things are being done. These qualities are more than latent capabilities; they also underscore the system's active role. If an institution embraces these qualities internally, its resource management efforts will reflect these beliefs.

Institutional conflict can be as damaging as external conflict, and the institutional system should provide mechanisms for dealing constructively with conflicts within and between its component parts and with other institutional systems. This process can be facilitated through the mutual learning that can occur in open policy dialogues in which the discovery of areas of both agreement and disagreement is valued. For a summary of these points, see Table 4.1 in the Appendix at the end of this chapter.

ENLARGED POLICY DEVELOPMENT CONTEXTS

Most of the ideas in this area are self-explanatory, but we wish to highlight those we think are most critical to the decision making that is fundamental to integrated resource management. In addition to institutional structures having the capacity for self-organizing around current and emerging issues, the capacity for proactive strategic policy intervention, and the ability to convene networks of transdisciplinary collaboration around these issues, it is necessary that they have the ability to close the feedback loops between resource managers at the local level (normally) and policy-makers at the provincial and federal levels. There is ample evidence of the truth of this. In the collapse of the cod fishery on the East Coast, for example, information that the hook-and-line fishers sought to communicate to federal officials was delayed in its transmission and then discounted. In addition, outside scientific expertise was not factored into the political decision making. The increasing interdependence of human and natural systems, coupled with the scale of human populations, makes it even more imperative that integrated resource management systems be established so that the people on the ground can communicate directly the changes they observe happening, as they are happening, without bureaucratic filters. This would occur if governments were to enlarge their policy-making contexts, as depicted in Figure 4.1 (Dale 2001), and actively integrate direct feedback from resource managers into the policy development process.

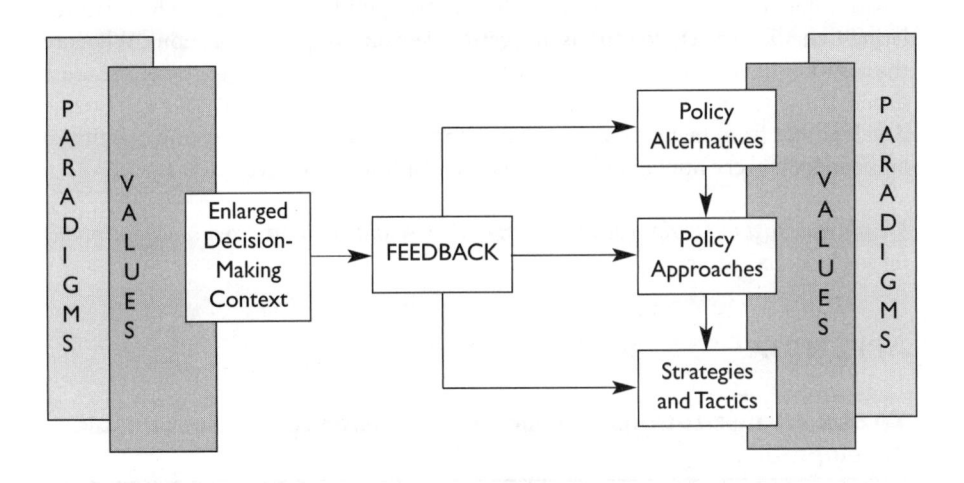

FIGURE 4.1 Enlarging the Context of Government Policy-Making

Reprinted with the permission of the publisher from 'At the Edge: Sustainable Development in the 21st Century' by Ann Dale, copyright University of British Columbia Press, 2001. All rights reserved by the publisher.

This enlarged policy development paradigm is based on a civics approach for governance (Nelson 1991) that goes beyond the simple functioning of governments to encompass loosely structured governance structures that can spontaneously emerge and self-organize around issues and domains of appreciation rather than around 'old' problems. Such a civics approach is a prerequisite for the kind of explicitly integrative, pluralistic, interactive approach to resource management that broadens the decision-making context to include human values and institutional contexts (McCarthy 2003). In this way, policy development for natural resource management becomes a process involving the active soliciting of feedback from the users and protectors of resources, including the wider scientific community who monitor and evaluate the effects of policy decision making. In the face of the complex, interactive, and unbounded high stakes and the often competing values that characterize natural resource management, this is the only common sense approach to both policy development and, hopefully, integrated decision making in the domain of appreciation. Indeed, as resources become more scarce, conflict over competing and multiple uses of the same resource appears to be increasing, making expanded policy-making contexts and enlarged decision-making forums even more timely and relevant.

In the final analysis, if we are to optimize the considerable gifts of our species, notably our sentiency, then our institutions must be sufficiently flexible to mimic those characteristics of natural systems that have worked for millennia to preserve their diversity and their resilience without human intervention. If we are to do more than survive, but rather flourish, we need institutions that embrace the integration of environmental imperatives, welcome diverse viewpoints, and are capable of stimulating Canadian society to start asking certain key philosophical questions: What are the costs of our adaptation? What is the nature of the limits of the biosphere? Are there limits on the scale of our activities? These are the questions that will take sustainable development from being a concept that is applied issue by issue to being one that is embedded in every human life and every facet of human society.

We are indebted to Dr Kai Lee for his email correspondence that provided the stimulus for some of these ideas.

APPENDIX

TABLE 4.1 Institutional characteristics that support sustainable development

Integrated and Coordinated	Reconciliation
Integrative: Each part of an institutional system interprets its mandate broadly to take into account all three dimensions of sustainability (social, economic, and ecological).	*Integrative:* All decision making for sustainable development fundamentally integrates ecological, social, and economic imperatives within a guiding framework.
Comprehensive: Each part of the institutional system recognizes all values associated with the resources it addresses and/or the services it delivers. It employs the principle of 'full cost accounting' in assessing the outcomes and impacts of decisions.	*Comprehensive:* Competing paradigms and conflicting world views are explicitly recognized and made transparent as part of the decision-making process. Multiplicity of perspectives and multiple contexts are venues for expression.

TABLE 4.1 Institutional characteristics that support sustainable development (cont'd)

Coordinated and Transactive: Each part of the institutional system recognizes linkages with other parts of the system, seeks to harmonize its activities and those of others, and promotes a coordinated approach to achieving overlapping activities.

Transcendent: Each part of the institutional system recognizes it is a part of a larger whole that furthers democracy through strengthening civil society.

Efficient and Effective

Efficient: Institutional system seeks to reduce overlaps and redundancies in the mandates and activities of its component parts; two or more parts of the system do not duplicate efforts. This criterion recognizes, however, that some degree of overlap is necessary to support integration and to ensure the 'robustness' of the system in being able to respond to unexpected events.

Flexible and Responsive

Flexible: Unnecessary overlap and duplication are eliminated through integrated decision making and the development and continual refinement of a guiding framework for operating across government.

Effective: Each part of the institutional system has a sufficient mandate and the required level of staff and resources to run processes, make decisions, implement results, and monitor or review outcomes as necessary to achieve its objectives. The operation of the system produces meaningful results from the perspective of those operating in the system, as well as from that of recipients of services provided by the system.

Effective: Resources are efficiently and effectively deployed to respond to emerging issues, particularly at the domain level.

Long Term and Adaptive

Strategic and Anticipatory: System is perceptive, looking for present and future opportunities and challenges. It establishes priorities to take action based on an assessment of the scope of impacts, irreversibility of decisions or actions, and urgency; in addition, it has the capacity to address short-term crises, undertake long-term planning and also anticipate and respond to issues that occur at 'in-between' speeds.

Long Term and Responsive

Strategic and Restorative: System responds equally to ecological, social, and economic feedbacks; in particular, it has the ability to recognize and respond in a timely fashion to negative feedback from ecological systems, particularly with respect to the loss of diversity at all scales.

TABLE 4.1 Institutional characteristics that support sustainable development (cont'd)

	Contextual: Systems of governance recognize and respond to the differing time, place, and scale phenomena of natural systems and human activity systems, recognizing that there are absolute limits on human activities imposed by the biosphere. Accordingly, decision-making systems reconcile an ecological framework of spatial boundaries with socio-political boundaries, taking into account the finite limits on place and scale imposed by the biosphere. A cautious approach is to live below those limits, rather than near or at the limits, in order to maximize resilience of all systems.
Reflexive and Adaptive: Institutional system has the capacity to keep up with changing values and knowledge and to review and improve decision-making processes. It has the mandate and tools required for self-evaluation and self-modification. It shows leadership in questioning not only the way things are done, but whether in fact the 'right' things are being done. This is not just a latent capability but an active role.	*Responsive:* Decision-making processes are enlarged policy-making contexts, transdisciplinary forums that bring together a multiplicity of stakeholders with relevant experiences to bear on the issue, particularly natural and social scientists, public policy practitioners, and the non-governmental community.
Open, Balanced, and Fair	**Open and Inclusive**
Representative: Each part of the system provides opportunities for all affected interests to be represented in processes, decisions, and actions.	*Equitable Access:* Involvement in decision making by the plurality of interests concerned is key. Diversity of representation in processes, decisions, and actions, plus employment of a multiplicity of approaches, is emphasized.
Equitable: System ensures that the costs and benefits of decision-making processes and their outcomes are distributed fairly among those affected and provides appeal mechanisms for those who feel that their interests have been overlooked or undetermined.	*Embeddedness:* Identification with our connectedness, as well as recognition of our being a small part of a larger grouping, provides a foundation for concerns for history, intergenerational and global equity, awareness of the needs of 'others' (Josselson 1996).

TABLE 4.1 Institutional characteristics that support sustainable development (cont'd)

Participatory and Collaborative: Institutional system provides opportunities for individuals and groups representing different interests to cooperate in decision making and take actions that affect their future while sharing responsibility for outcomes.	*Networks of Collaboration:* Institutional system recognizes the complexity, incomplete knowledge, and uncertainty inherent in living systems, and the fact that no one sector can solve the complex societal issues involved. Its role is to stimulate networks of collaboration around 'domains of interest' leading to solutions and concrete actions.
Responsive and Accountable: Each part of the system responds in a timely fashion to the constituency it serves and provides mechanisms by which individuals or groups can be held responsible for a decision or action directly by that constituency; these mechanisms are not so rigid as to inhibit creativity.	*Integrity:* Integrity of information is critical to the responsiveness of the system, particularly negative feedback from ecological systems. In order for the system to be able to respond to negative feedback information, subsidiarity is fundamental.
Conflict-Resilient: System provides mechanisms to deal constructively with conflicts within and between its component parts and with other institutional systems.	*Open:* Mutual learning occurs in open policy dialogues that value discovering main areas of both agreement and disagreement.

SOURCES: Adapted from Rueggeberg and Griggs 1993; Dale 2001. Reprinted with permission of the publisher from 'At the Edge: Sustainable Development in the 21st Century' by Ann Dale, copyright University of British Columbia Press, 2001. All rights reserved by the publisher.

REFERENCES

Alabo, G., L. Languille, and L. Panitch. 1993. *A different kind of state: Popular power and democratic administration.* Don Mills, ON: Oxford University Press.

Bhaskar, R. 1994. *Plato, etc: The problems of philosophy and their resolution.* New York: Verso.

Bradford, N. 2003. *Why cities matter: Policy research perspectives for Canada.* Canadian Policy Research Networks discussion paper.

Bregha, F., J. Benidickson, D. Gamble, T. Shillington, and E. Weick. 1990. *The integration of environmental considerations into government policy.* Report prepared for the Canadian Environmental Assessment Research Council. Ottawa: Minister of Supply and Services Canada.

Brundtland, G.H. 1987. *Our common future.* Oxford: World Commission on Environment and Development.

Caldwell, L.K. 1974. Environmental policy as a catalyst of institutional change. *American Behavioral Scientist* 17 (5): 711–30 (May/June).

Carpenter, S.R., W.A. Brock, and D. Ludwig. 2001. Collapse, learning and renewal. In L. Gunderson, C.S. Holling (eds), *Panarchy: Understanding transformations in human and natural systems.* Washington, DC: Island Press.

CPRN (Canadian Policy Research Networks). 2003. *A citizens' dialogue on Canada's future.* Ottawa: Internal publication.

Dale, A. 2001. *At the edge: Sustainable development in the 21st century.* Vancouver: University of British Columbia Press.

——— 2005. Social capital and sustainable community development: Is there a relationship? In A. Dale and J. Onyx, *A dynamic balance: Social capital and sustainable community development.* Vancouver: University of British Columbia Press.

Dietz, T., and T. Burns. 1992. Human agency and the evolutionary dynamics of culture. *Acta Sociologica* 35:187–200.

Dobell, R. 1995. The 'dance of the deficit' and the real world of wealth: Re-thinking economic management for social response. Paper prepared for the National Forum on Family Security, March.

Emery, F.E., and E.L. Trist. 1972. *Towards a social ecology. Contextual appreciation of the future in the present.* London: Plenum Press.

Gunderson, L.H., and C.S Holling (eds). 2002. *Panarchy: Understanding transformations in human and natural systems.* Washington, DC: Island Press.

Harvey, D. 2002. Agency and community: A critical realist paradigm. *Journal for the Theory of Social Behavior* 32 (2): 163–94.

Hill, S.B. 2005. Social ecology as a framework for understanding and working with social capital and sustainability within rural communities. In A. Dale and J. Onyx, *A dynamic balance: Social capital and sustainable community development.* Vancouver: University of British Columbia Press.

Hoberg, G. 1992. *Pluralism by design: Environmental policy and the American regulatory state.* New York: Praeger.

Holling, C.S. 2003. From complex regions to complex worlds. Starker Lecture Series, Oregon State University.

Holling, C.S., L.G. Gunderson, and G.D. Peterson. 2002. Sustainability and panarchies. In L.H. Gunderson and C.S. Holling (eds), *Panarchy: Understanding transformations in human and natural systems,* 63–102. Washington, DC: Island Press.

Horlick-Jones, T, and J. Sime. 2004. Living on the border: Knowledge, risk, and transdisciplinarity. *Futures* 36:441–56.

Inglehart, R. 1990. *Culture shift in advanced industrial society.* Princeton: Princeton University Press.

Jacobs, J. 2004. *Dark age ahead.* Toronto: Random House.

Josselson, R. 1996. *The space between us: Exploring the dimensions of human relationships.* Thousand Oaks, CA: Sage Publications.

Lesser, E., and L. Prusak. 2000.Communities of practice, social capital and organizational knowledge. In E. Lesser, M. Fontaine, and J. Slusher (eds), *Knowledge and communities.* Boston: Butterworth Heinemann.

McCarthy, D.D. 2003. Post-normal governance: An emerging counter-proposal. *Environments* 31 (1): 79–87.

Macdonald Commission. 1985. *Royal Commission on the Economic Union and Development Prospects for Canada.* 3 vols. Ottawa: Minister of Supply and Services Canada.

McFarland, J. 2003. Good governance group set. *Globe and Mail*, 12 April.

Mortsch, L.D., and B. Mills (eds). 1996. *Great Lakes–St. Lawrence Basin Project on Adapting to the Impacts of Climate Change and Variability—Progress report one*. Ottawa: Environment Canada.

Nelson, J.G. 1991. Research in human ecology and planning: An interactive, adaptive approach. *Canadian Geographer* 35 (2): 114–27.

OECD (Organization for Economic Co-operation and Development). 2002. Improving policy coherence and integration for sustainable development: A checklist. *OECD Observer*, October.

Pross, P.A. 1992. *Group politics and public policy*. 2nd edn. Don Mills, ON: Oxford University Press.

Putnam, R. 2000. *Bowling alone: The collapse and revival of American community*. New York: Simon & Schuster.

Ramadier, T. 2004. Transdisciplinarity and its challenges: The case of urban studies. *Futures* 36:423–39.

Robinson, J.B., and J. Tinker. 1997. Reconciling ecological, economic and social imperatives: A new conceptual framework. In T. Schrecker (ed.), *Surviving globalism: Social and environmental dimensions*. London: Macmillan.

Rothman, D., and J.B. Robinson. 1997. Growing pains: A conceptual framework for considering integrated assessments. *Environmental Monitoring and Assessment* 46:23–43.

Rueggeberg, H., and J. Griggs. 1993. Institutional characteristics which support sustainability. In S. Peck (ed.), *1995 Environmental Scan for the Canadian Council of Ministers of the Environment*. Toronto: Thompson, Gow and Associates.

Shackley, S., and B. Wynne. 1995. Integrating knowledges for climate change: Pyramids, nets and uncertainties. *Global Environmental Change* 5 (2): 113–26.

UNDP (United Nations Development Programme). 1997. *Governance for Sustainable Human Development: A UNDP Policy Document*. http://mirror.undp.org/magnet/policy/.

Versteeg, H. 1992. *A case study in multi-stakeholder consultation: The corporate history of the Federal Pesticide Registration Review*. 2 vols. Ottawa: Canadian Centre for Management Development.

Chapter 5

Integration through Sustainability Assessment: Emerging Possibilities at the Leading Edge of Environmental Assessment

Robert B. Gibson

INTRODUCTION

Integrated resource management and environmental assessment grew up in different neighbourhoods, went to different schools, and found employment in different agencies. While the roots of integrated resource management lie in the failures of conventional forestry and fisheries administration, environmental assessment was meant to correct inadequacies in pollution control and environmental protection regimes. Nevertheless, their basic job descriptions have been more or less the same. Both have been expected to deal with a world that is too complex, too uncertain, and too demanding to be left to the simple mechanism of the market or the narrow mandates of established authorities.

After some decades of effort and evolution, neither integrated resource management nor environmental assessment has emerged as the salvation of the world. Successes can be claimed, exemplary applications identified, and significant advances celebrated, but perhaps the greatest accomplishments of the last three decades or so of integrated resource management and environmental assessment have been a richer understanding of complexity, a deeper foundation for uncertainty, and an expanded set of demands for vision and inclusion.

Some may find this depressing. Certainly it is unhappy news for those who still hope for managerial rationality and administrative convenience. There is, however, a cheerful aspect. It is the gradual emergence of new approaches to deliberation and decision making—on resource management and a host of other practical challenges—that are suitable for a world of complexity, uncertainty, and multiple demands. Some of these have been discussed in earlier chapters of this book. This chapter is devoted to the nature and implications of *sustainability assessment*, a closely related but more comprehensive, more versatile, and potentially more effectively integrative approach that is arising concurrently in many venues, but most visibly now on the advanced edges of environmental assessment.

Sustainability assessment is perhaps as much the progeny of innovative and responsive practice in land-use planning, urban design, development assistance, and resource management as it is of environmental assessment. What is being learned in environmental assessment owes much to insights borrowed from elsewhere. And as a larger package, sustainability assessment is more accurately a synthesis of these various influences and components than a simple expansion of environmental assessment. But perhaps the essential character of the idea is most easily illustrated in the maturing of environmental assessment—in its transition from the modest aims and fragmented components of conventional environmental impact assessment to the higher-test, longer-view, and more comprehensive agenda of sustainability-based processes.

As we will see, environmental assessment has helped force some consideration of usually neglected biophysical and socio-economic factors in conventional decision making. But in most applications, including those directly relevant to resource management, consideration of environmental effects has not ensured very effective integration of these concerns in the planning, design, and implementation of significant undertakings.

There is no guarantee that a transition to sustainability assessment will ensure effective integration of considerations either. But adoption of the sustainability test does make assessment more comprehensive and should situate it at the core of decision making. The challenge to be addressed here is how to design sustainability assessment obligations so that they can be vehicles for effective integration.

THE EVOLUTION OF ENVIRONMENTAL ASSESSMENT

Over the past 30-some years many jurisdictions around the world have chosen to impose environmental assessment requirements on those seeking approval for undertakings that may have significant negative effects. The requirements have not always been firmly mandatory, and the specifics—the range of undertakings subject to assessment, the procedural steps, the scope of considerations, the nature and roles of participants, the flexibility of application, and even the criteria to be satisfied—have varied widely. But the essential purpose has remained: environmental assessment is meant to change the nature of decision making. Under ordinary circumstances, decision makers can be relied upon to consider economic and technical matters, at least those tied to their immediate objectives and mandates. Governments also have reliable incentives to pay attention to political concerns. Environmental assessment was introduced as an obligation as well as a means to take environmental factors just as seriously.

Like the expansion of environmental protection regulations that began earlier and continued over the same period, environmental assessment requirements came as government responses to public concerns about the negative effects of undertakings that had gone ahead without serious regard for potential biophysical and socio-economic impacts. Because of this foundation in public concern, environmental assessment processes have usually been seen as means of improving the openness as well as the environmental responsibility of decision making.

Although environmental assessment emerged from the same pressures as regulatory initiatives, it represented a more ambitious attempt to reform decision making. Environmental regulations have required identification of certain kinds of environmental impacts, but have typically focused on the activities of individual sectors (e.g., projects of the nuclear industry) or particular receptors (e.g., damage to fish habitat or pollution of air) and have attempted to ensure compliance with specified standards. While useful, these initiatives have tended to be too reactive, too narrowly focused, and too tolerant of reduced but continuing degradation. Environmental assessment requirements represent an improvement insofar as they apply earlier and more broadly in the process, encouraging proponents and decision makers to address environmental concerns from the outset of planning. As well, environmental assessment provides for consideration of biophysical and (often) socio-economic effects in a more comprehensive and, at least potentially, better-integrated package.

Generally, proponents subject to environmental assessment requirements are required to carry out specific research on the existing environment and how it could be affected by proposed undertakings. To ensure that the research findings are valid and are used properly in the choice and detailed design of proposals, proponents are required to seek approval through a formal review and decision-making process. Public involvement and consultation with relevant government agencies are encouraged if not mandatory in most processes. Usually, there are more or less formal mechanisms for incorporating assessment review conclusions in approval and implementation decisions.

Within this very loose framework, however, the particulars have varied greatly and in important ways. Given the purpose of ensuring serious attention to environmental considerations, it matters a great deal whether environmental assessment is introduced through voluntary encouragement or through legal mandate and whether *environment* is defined narrowly to cover only biophysical and ecological matters or is extended to cover socio-economic and cultural concerns as well. It matters whether proponents of new projects begin their deliberations knowing they will have to satisfy environmental assessment requirements or whether they have more limited expectations imposed after they have decided what they wish to do. It matters whether the process is efficient or cumbersome, whether the joint roles of science and preference are understood, and whether those with the most to lose can get a fair hearing. On all these topics and more, the theoreticians and practitioners of environmental assessment have now struggled for more than 30 years. Considerable diversity remains—in thinking as well as in practice. But it is now possible to look back on the birth and growth of environmental assessment and to identify the key components and most significant areas of maturation.

The two columns in Table 5.1 summarize the main components of basic assessment processes and list the major categories of additional components that have been adopted in more advanced regimes. Not all environmental assessment regimes yet incorporate the full set of basic components. Some authorities with initially strong processes have chosen to soften requirements or restrict application in the name of greater efficiency or devotion to immediate economic gains. But the overall trend

TABLE 5.1 The structure and key components of potentially effective assessment processes

Basic Components	Additional Components of More Advanced Processes
1. Application rules that specify what sorts of undertakings are subject to assessment requirements (so planners and proponents know from the outset that they will have to address environmental considerations)	1. Application rules that ensure assessment of all undertakings, including policies and programs and plans as well as capital projects, that might have significant environmental effects
2. Guidance and procedures for determining more specifically the level of assessment and review required in particular cases	2. Requirements to establish the need and/or justify the purpose to be served
3. Definition of the range of 'environmental' considerations to be addressed, preferably including socio-economic and cultural as well as biophysical factors	3. Requirements to identify the reasonable alternatives, including different general approaches as well as different designs, for serving the purpose
4. Requirements to identify and evaluate the potentially significant effects of proposed undertakings, in light of existing environmental conditions, pressures, and trends	4. Requirements for integrated consideration of related undertakings and of cumulative effects of existing, proposed, and reasonably anticipated undertakings
5. Provisions for scoping (setting reasonable boundaries and focusing assessment work on the most important issues)	5. Requirements to identify means of enhancing positive effects
6. Requirements to identify and evaluate means of mitigating predicted negative effects	6. Requirements for comparative evaluation of the reasonable alternatives, with justification for selection of the preferred alternative as the proposed undertaking
7. Overall evaluation of the effects of the proposed undertaking, with chosen mitigation measures	7. Requirements to identify and evaluate the significance of uncertainties (about effect predictions, mitigation, and enhancement effectiveness) and associated risks
8. Provisions for public as well as technical review of the proposed undertaking and the assessment work (to evaluate both the proposed undertaking and the adequacy of efforts to incorporate attention to environmental considerations in developing the proposal), including review through public hearings in especially significant cases	8. Provisions, including funding support, to ensure effective public as well as technical notification and consultation at significant points throughout the proposal development and assessment process
9. Means of ensuring that the assessment and review findings are incorporated effectively in approvals and permitting	9. Requirements and provisions for monitoring of actual effects and comparison of these with predicted effects (to allow adaptive management and enhance learning from experience) through the full life cycle of the undertaking

TABLE 5.1 The structure and key components of potentially effective assessment processes (cont'd)

Basic Components	Additional Components of More Advanced Processes
10. Requirements and provisions for monitoringing and enforcing compliance with approval conditions	10. Provisions for linking assessment work, including monitoring, to a broader regime for setting, pursuing, and re-evaluating public objectives

SOURCES: Adapted from Gibson 1993 and 2002b, with insights from Sadler 1996; CSA 1999; Senécal et al. 1999; Wood 2003; Lawrence 2004; and André et al. 2004.

has been towards more widespread and consistent adoption of the more advanced components.

These changes reflect an evolution in thinking and practice over the years that has responded both to lessons from assessment experience and to emerging understandings about how to manage human activities generally. Several of the shifts are the result of pressures to make environmental assessment more influential. These include steps to make assessment

- more mandatory and codified (increased adoption of law-based processes, further specification of requirements, reduction of discretionary provisions);
- more widely applied (covering small as well as large capital projects, continuing as well as new initiatives, sectoral and area developments as well as single proposals, strategic- as well as project-level undertakings);
- more open and participatory (involving a broad range of civil society organizations and local residents as well as just proponents, government officials, and technical experts); and
- more closely monitored (by the courts, informed civil society bodies, and government auditors watching responses to assessment obligations, and by stakeholders watching actual effects of approved undertakings).

For our purposes here, the more significant changes may be those centred on the substance of assessment practice—the adoption of approaches that are better suited to a world that behaves as a set of deeply intertwined complex systems and in which human activities are imposing increasingly unsustainable demands. To deal with these realities, environmental assessment has been evolving to be

- more critical (more often initiated early in planning, considering purposes and broad alternatives, sometimes beginning with the driving policies, programs, and plans);
- more comprehensive of environmental concerns (socio-economic, cultural, and community effects; biophysical and ecological effects; as well as regional, global, and local effects) and their interrelations (attention to cumulative and systemic effects rather than just individual impacts);

- more accepting of different kinds of knowledge and analysis (informal and traditional knowledge as well as conventional science; preferences as well as 'facts');
- more sensitive to complexity and uncertainties (requiring confidence estimates, applying precaution);
- more often adopted beyond formal environmental assessment processes (through sectoral law at various levels, but also in land-use planning, through voluntary corporate initiatives, etc.); and
- more ambitious (aiming for overall biophysical and socio-economic gains rather than just individually 'acceptable' undertakings).

These changes have been complemented by many other pressures on public and private sector proponents to take environmental concerns seriously (liability worries, regulatory due diligence expectations, consumer expectations, insurance and lender demands, etc.) in a wider range of activities. And while some jurisdictions and corporations have demonstrated more commitment than others, most are now giving environmental concerns much more extensive and serious attention than they did when environmental assessment was first introduced. Unfortunately, extensive and serious attention is not the same as well-integrated and effective influence.

ENVIRONMENTAL ASSESSMENT AS A TOOL FOR INTEGRATION

In the early years, environmental assessment advocates hoped that information would be enough. The US National Environmental Policy Act of 1969 was satisfied with requiring proponents to carry out impact assessment studies, and relied on the relevant decision makers to take the findings into account. Many other subsequent regimes, including successive versions of the Canadian federal process, similarly treated environmental assessment as a source of additional considerations for decision makers to incorporate in their otherwise conventional approvals.

The hope was not just that the authorities, in their wisdom, would make use of the information, but also that attention to environmental considerations would gradually become institutionalized as a regular component of deliberations on relevant undertakings (André et al. 2004, 101ff). To some extent this has happened. Public and private sector bodies that regularly propose environmentally significant projects now typically have internal staff with expertise and mandates for addressing environmental considerations (or for directing the external consultants who do much of the actual assessment work). Proponent agencies and corporations also have established decision-making processes that recognize assessment requirements and their potential influence on design, scheduling, costs, and other core issues. Moreover, assessment work is now commonly linked to continuing environmental management efforts and to the operation of environmental management systems where these are in place (Wood 2003, 6).

Nevertheless environmental assessment is still not well integrated into the core of decision making on major undertakings. Many proponents continue to treat environmental assessment as a marginal approval requirement—an irritating obligation to jump through yet another administrative hoop—and complain bitterly if the timing and conclusions are not firmly predictable. Many approval authorities similarly treat environmental assessment as an added burden tangential to their main concerns. They resist the establishment of independent assessment agencies with approval powers and keep final decision making to themselves, outside the ambit and scrutiny of public assessment processes.

The evolution of environmental assessment processes is, however, moving slowly to push for more open and effective integration. Indeed, this is happening at four levels: impact assessment work, project decision making, assessment of strategic undertakings, and links beyond assessment processes.

Integration in Impact Assessment Work

Primitive environmental impact studies were, and are, condemned for providing inventory lists rather than descriptions of key components and relationships in the existing environment and for offering vague estimates of direct effects on particular receptors instead of testable predictions of overall impacts. Many assessment professionals and participants, most notably Beanlands and Duinker (1983), argued for ecosystem-based approaches focused on the components most valued by experts and citizens. Their recommendations have been adopted slowly. Implementation has only gradually spread into consideration of valued socio-economic and cultural components, and further integration of the social and ecological spheres has been hesitant. But impact assessment practice has benefited from the advancement of learning about complex systems combined with assessment process expansion to address the interrelations of social, economic, and cultural, as well as biophysical, effects.

Integration in Project Decision Making

As noted above, few authorities have been inclined to see environmental assessment as a core element of decision making. Many assessments have been treated as an omnibus approach to environmental regulatory permitting. And where assessment findings have raised unavoidably important implications for overall approvals, the processes for incorporating these considerations with other more conventional financial, technical, and political concerns have typically been shrouded in secrecy.

This deficiency is exacerbated by the habit of designing assessment processes that concentrate on informing the project approval decision. The project approval—the key ruling on whether the proposed undertaking should go ahead, and if so, under what terms and conditions—is an understandable but misleading focus. Approval decisions are important, visible, and common to most undertakings, but the usual project reality is a long series of successive, sometimes iterative deliberations and choices over the full life of an undertaking, from initial conception to final decommissioning. The most significant decisions are often the early ones on needs, purposes, and alternatives. Post-approval arrangements for monitoring, enforcement, adaptation, and review may also have important consequences. Limiting attention to the project approval sharply reduces the potential benefits of the exercise.

Effective integration of assessment findings in the full suite of project decisions is much more common in advanced regimes, where assessments examine basic needs and purposes, cover the full suite of socio-economic and biophysical effects, compare alternatives, and seek to identify best options, and where means of facilitating effective public involvement are in place. But even these provide little guidance on how the environmental considerations are to be integrated with other priorities.

Integration through Strategic Assessment

Assessment critics have long noted that the traditional project-centred regimes tend to overlook our most significant environmental worries. While an individual project focus is convenient (identifiable proponents, clear decision points, and specified approval mandates), the reality of impacts is that they come from multiple undertakings and activities. Moreover, the most important determinants of impacts may not be the individual projects, or even groupings of them, but rather the policies, programs, plans, regulatory initiatives, and tax rulings that set the effective context of incentive and guidance.

Here, too, advanced assessment regimes have responded. They have introduced mandatory attention to cumulative effects and have extended application to the strategic level of plans, programs, and policies. Sometimes this has been accomplished by initiatives outside environmental assessment proper (e.g., in planning reforms, commissions of inquiry, innovative resource management regimes, and environmental auditing bodies), but the effect is similar.

In all cases, the value of the initiatives is heavily dependent on how well the strategic- and project-level deliberations and findings are integrated. Sometimes strategic assessments are used to provide a set of substantive rules for subsequent more particular undertakings (e.g., where regional planning establishes the framework into which municipal plans and development decisions must fit). Sometimes the main strategic contribution is a defined process for more specific planning or project development. Too often, however, the strategic-level work provides neither framework nor process, just information and broad guidance that may be considered but have no clear official status.[1]

Integration beyond Assessment Processes

At best, environmental assessment at the strategic and project levels can only be a contributor to the pressures, processes, and practices needed to foster more forward-looking and comprehensively informed decisions. Among the other requirements are good indicators of present conditions and trends, scenarios of desirable and viable futures, sustainability-oriented education and research, appropriate regulatory and fiscal regimes, participative monitoring, and a habit of regular review, reconsideration, and adjustment. Fixed formal links among all these would probably serve poorly. The whole must be flexible and responsive. But integration at least in the sense of shared enlightenment and mutual reinforcement would be welcome, perhaps necessary.

No such arrangements are in place in any existing jurisdiction, though many of the pieces can be identified and in many places there has been gradual movement towards better integration of governance systems and more serious attention to sustainability commitments. Advances in environmental assessment processes have contributed to this, and further steps will be enabled by expanded adoption of a sustainability-centred agenda for assessment processes.

MATURING ASSESSMENT AND EMERGING SUSTAINABILITY

The trends discussed above are leading environmental assessment into a bigger range of application (from the strategic to the project level), with a broader agenda (attention to purposes, alternatives, and a full suite of interrelated effects, individual and cumulative), more sophisticated understanding (systemic and precautionary), more players (civil society organizations and traditional knowledge holders), and higher ambitions.

None of this movement has been accidental. Each of the shifts has been the product of concerted effort in the face of often stiff resistance and with continuing tensions. In some areas, progress along the trend line has been modest and tentative. And retreats have sometimes accompanied the overall advances. However, the story here has not just been one of struggles among competing interests. The trends also reflect a response to realities that have become more evident or more pressing in recent decades and that decision makers ignore at their peril. As noted above, these realities include the rise and persistence of sustainability as a fundamental concern and objective.

Sustainability is a difficult concept, not yet well elaborated for assessment purposes. But it clearly involves a combination of aspects that overlap closely with the evident trends in environmental assessment maturation. Any planning, decision, and follow-up process that aims to contribute to sustainability must surely be comprehensive and integrative, critically attentive to purposes and alternatives, appreciative of uncertainties, and applied firmly, widely, openly, and efficiently.

It does not follow that assessment processes with these characteristics will necessarily serve well as well-integrated vehicles for the pursuit of sustainability. It is one thing to be sufficiently broad and ambitious and also have suitable participation and humility. It is quite another to be clear about what is needed for reasonable progress towards sustainability, what improvements are crucial, what compromises can be tolerated, generally and in specific circumstances, and how all this can be fitted smoothly with the range of other needed initiatives. For that, we need to look much more closely at what sustainability assessment might entail.

THE RISE OF SUSTAINABILITY ASSESSMENT

Sustainability assessment initiatives are proliferating quickly around the world. At the April 2004 annual conference of the International Association for Impact Assessment, dozens of paper presenters reported on efforts to apply some form of sustainability analysis, appraisal, or assessment or otherwise to adopt sustainability objectives as core guides for evaluations and decisions. Case applications covered a host of different proponents, scales, and activities, including offshore gas development in Western Australia (Pope 2004), mining projects in Labrador and northern British Columbia (Hodge and Thomson 2004), an urban sustainability agenda in Graz, Austria (Aschemann 2004), winter Olympic games preparations in Vancouver (Bekhuys and McKay 2004), and district plans for implementing Ghana's poverty reduction strategy (Nelson et al. 2004). Moreover, they were backed by a growing library of broader explorations covering applications in whole sectors, scales, and regions.[2]

These initiatives represent just a small sampling of what is underway. Any capable Internet search engine will now uncover hundreds of government, corporate, academic, civil society, even personal websites presenting work labelled as sustainability assessment. And these sustainability assessment efforts are accompanied by an even more extensive set of evidently serious attempts to define sustainability objectives, to identify appropriate indicators, and to apply sustainability criteria in important decision making, public and private, at all levels, from the local to the global.

The vast diversity of sustainability assessment experiments includes many with tenuous claims to the category. As in the larger realm of asserted commitments to sustainable development, conceptual rigour and effective action are much less common than cheerful visions and passionate endorsements. Nevertheless, the great proliferation of sustainability assessment initiatives is clearly a response to widespread and genuine pressures for more effectively comprehensive, far-sighted, critical, and integrated approaches to project- and strategic-level decision making.

These pressures are likely to increase. Citizens and authorities are increasingly aware of the interconnections among economic, social, and ecological considerations. The costs and perils of unsustainable behaviour are becoming more evident at every level. Authorities who have now spent well over a decade making formal commitments to sustainability are being pressed to act accordingly. And, after lengthy contests over the meaning of *sustainability* and *sustainable development*, there is some emerging consensus on the fundamentals.

To date, no sustainability assessment initiative provides an ideal model. Nevertheless, it is possible now to sketch out the main components of a viable approach:

- the basic sustainability requirements that inform a transition to sustainability assessment;
- the main implications of these requirements for sustainability assessment processes; and
- possible approaches to the most challenging areas, including how to define such core sustainability requirements as evaluation and decision-making criteria and how to deal with the inevitable compromises and trade-offs among these requirements.

THE FOUNDATIONS OF SUSTAINABILITY ASSESSMENT
Sustainability Concept Basics

Sustainability is a very old idea. Except for the last few hundred years, most societies other than those engaged in empire building were essentially customary; they were chiefly concerned with maintaining and continuing well-tested ways of doing things. Recent attention to sustainability is of a different sort. It presumes a world of change and seeks progress, though of a different sort from what now prevails. Essentially, the present concept of sustainability stands as a critique. It is a response to evidence that current conditions and trends are not viable in the long run and that the reasons for this are as much social and economic as they are biophysical or ecological.

Since 1987, when the World Commission on Environment and Development issued its report, *Our Common Future*, the terms *sustainability* and *sustainable development* have been widely, if sometimes cynically, embraced by public and private sector bodies. There has been much debate about the meaning and implications of serious commitment to sustainability, and these deliberations continue. But after a decade and a half of experimentation as well as study, there has been evident progress towards consensus on the fundamentals, supported by complementary developments in several adjacent areas of theory and practice.[3]

The following six points are now safe assertions about the conceptual essentials for pursuing sustainability, at least for the purposes of sustainability assessment:

- Sustainability considerations include socio-economic as well as biophysical matters and are especially concerned with the interrelations between and interdependency of the two.
- Human as well as ecological effects must be addressed as parts of large complex systems.
- Because the complexity of these systems makes full description impossible, prediction of changes uncertain, and surprise likely, precaution is needed.
- Minimization of negative effects is not enough; assessment requirements must encourage positive steps towards greater community and ecological sustainability and towards a future that is more viable, pleasant, and secure.
- Corrective actions must be woven together to serve multiple objectives and to seek positive feedbacks in complex systems.
- While a limited set of fundamental, broadly applicable requirements for progress towards sustainability may be identified, many key considerations will be location specific, dependent on the particulars of local ecosystems, institutional capacities, public preferences, etc.

These points have implications for how we should design assessment processes as well as for how we should make judgments within the processes (about what undertakings should be assessed, what effects are significant, what alternatives should be preferred, what compromises can be accepted, etc.). The process implications are particularly important because many of the key practical considerations are case specific. Different places face different stresses, offer different opportunities, and merit different priorities. Providing for case-specific deliberations and judgments is a crucial feature of assessment processes.

At the same time, a solid, if broad, understanding of the key requirements for progress towards sustainability is needed as a basic set of criteria both for the decisions to be made in process design and for elaboration of the more case-specific criteria needed for judgments in particular assessments. After all, the intent is for the many individual, situation-specific decisions to contribute to an overall transition to behaviour that is on all fronts increasingly sustainable.

Here, too, the fundamentals can be set out without much difficulty. A sizable portion of the voluminous literature on sustainability seeks to define the basics of sustainable development or sustainability in and through universally applicable principles and criteria. The results include a wide spectrum of views on the proper organizing categories. For some years there were lively debates about whether it is best to conceive of sustainability as resting on two intersecting pillars (the ecological and the human) or on three (social, ecological, and economic) or five (ecological, economic, political, social, and cultural) or more.[4] Others have found non-pillar or cross-pillar approaches to be better suited to the intersecting character of sustainability requirements.[5] But all this has been essentially about emphasis. There is broad agreement on the key considerations.

Building on the key elements of agreement from the last two decades of thinking and experimentation on sustainability and its application, Table 5.2 presents the basic sustainability requirements as a set of decision criteria for application in sustainability assessments. In the interests of fostering better-integrated thinking and analysis, this list carefully avoids the pillars, though it incorporates the key concerns from the usual pillar-based categories (ecology, economy, society, etc.). In the early days of sustainability discussions, identifying the pillars helped to underscore that all of them must be involved in efforts to secure a durable and desirable future. But defining sustainability needs in the familiar but separate categories of ecology, politics, society, economics, and culture perpetuates fragmentation. Most participating individuals and agencies already come to the sustainability assessment table with particular areas of expertise, mandate, and interest to apply and defend. Encouraging them to think and act outside these boxes is easier when sustainability is defined in ways that stress the interconnections and go more directly to the substance of what must be considered and done. Thus, the list in Table 5.2 is designed to concentrate attention on what must be achieved—and what key actions are involved—to move consistently towards greater sustainability.

While the set of general decision criteria in Table 5.2 is broadly comprehensive and friendly to integration, it is just a working list of the titles of general requirements. Although these requirements are based on a careful synthesis of literature and case experience (Gibson et al. 2005), there is no reason to insist on this particular formulation. The items could be subdivided, reconstructed, reordered, and reworded in a host of different ways. And like any such offering, this list is properly subject to continued learning and adjustment. More importantly, Table 5.2 only sets out the general requirements. The specifics of each item and the package as a whole must be defined

TABLE 5.2 General sustainability requirements as criteria for sustainability assessment decision making

Socio-ecological system integrity

Build human-ecological relations to establish and maintain the long-term integrity of socio-biophysical systems and protect the irreplaceable life-support functions upon which human as well as ecological well-being depends.

Sufficiency and opportunity

Ensure that everyone and every community has enough for a decent life and that everyone has opportunities to seek improvements in ways that do not compromise future generations' possibilities for sufficiency and opportunity.

Intragenerational equity

Ensure that sufficiency and effective choices for all are pursued in ways that reduce dangerous gaps in sufficiency and opportunity (and health, security, social recognition, political influence, etc.) between the rich and the poor.

Intergenerational equity

Favour present options and actions that are most likely to preserve or enhance the opportunities and capabilities of future generations to live sustainably.

Efficiency

Provide a larger base for ensuring sustainable livelihoods for all while reducing threats to the long-term integrity of socio-ecological systems by reducing extractive damage, avoiding waste, and cutting overall material and energy use per unit of benefit.

Democracy and civility

Build the capacity, motivation, and habitual inclination of individuals, communities, and other collective decision-making bodies to apply sustainability requirements through more open and better-informed deliberations, greater attention to fostering reciprocal awareness and collective responsibility, and more integrated use of administrative, market, customary, and personal decision-making practices.

Precaution and adaptation

Respect uncertainty, avoid even poorly understood risks of serious or irreversible damage to the foundations for sustainability, plan to learn, design for surprise, and manage for adaptation.

Immediate and long-term integration

Attempt to meet all requirements of sustainability together as a set of interdependent parts, seeking mutually supportive benefits.

SOURCES: From Gibson 2002a and Gibson et al. 2005, chap. 5

in context by the relevant communities of interest and concern. How this specification is done—what processes are used for the discussions and choices involved, how the means fit with the ends—is no less important than the general requirements to be respected.

The substance of this list of core sustainability requirements does, however, have some clear and immediate implications. Some of these inform its role as a guide for designing sustainability assessment processes.

SUSTAINABILITY ASSESSMENT PROCESS BASICS

Sustainability-based assessment processes are not the only vehicles for the pursuit of sustainability. But because of their potential for influencing the preparation, evaluation, approval, and implementation of a wide range of significant undertakings—policies, plans, and programs as well as projects—they are particularly well suited to the task. Advanced environmental assessment processes of the kind outlined in Table 5.1 above provide a solid foundation. They are characteristically anticipatory and forward looking, integrative, flexible enough for application to very different cases in very different circumstances, generally intended to force attention to otherwise neglected considerations, open to public involvement, and adaptable in ways that suggest capacity for progressive evolution.

As we have seen, the key deficiencies of existing environmental assessment processes include integration inadequacies. Few environmental assessment processes today are well designed for addressing human and ecological effects together within complex systems. Few emphasize attention to maximizing overall positive long-term improvements. Most fail to ensure effective integration of environmental considerations in the key early decisions on purposes and preferred options. And too often the results are merely advisory, have little influence in final decisions, or are incorporated with compromises and trade-offs that are reached through separate, non-transparent negotiations wherein environmental matters are still treated as constraints, in conflict with priority objectives.

Adoption of a sustainability-centred focus will not automatically overcome these deficiencies. But sustainability is an essentially integrative concept. Its essentials recognize the interdependence of social, economic, cultural, political, and biophysical factors; cover the interests of future as well as present generations; apply to the full range of human undertakings; and are concerned with positive as well as negative effects in the interests of maximizing net gains. Combined with the recognized components of advanced environmental assessment processes, the sustainability commitment encourages much more effective integration.

Sustainability assessment processes that are built on the general model of progressive environmental assessment regimes (integrating strategic- as well as project-level processes) promise three major changes. First, they would force attention to sustainability requirements and would identify and judge trade-offs in light of a commitment to making positive contributions to achieving sustainability objectives. Second, sustainability assessment would specify these requirements and trade-off judgments—and associated values, objectives, and criteria—in specific contexts, through informed choices by the relevant parties (stakeholders). Finally, sustainability assessment would apply these insights in the full set of process elements recognized in progressive environmental assessment processes:

- requiring special attention in the planning and implementation of significant undertakings at the strategic and project levels;
- identifying appropriate purposes and options for new or continuing undertakings;
- assessing purposes, options, impacts, mitigation, and enhancement possibilities, etc.;
- choosing (or advising decision makers on) what should (or should not) be approved and done, and under what conditions; and
- monitoring, learning from the results, and making suitable adjustments during implementation.

The basic design features for sustainability assessment processes are not significantly different from those for strong environmental assessment regimes. Adjusted for the sustainability mandate, the main assessment process qualities are those set out in Table 5.3. No existing jurisdiction has incorporated all of these features in the design and implementation of assessment processes, even for more limited environmental assessment objectives, and sustainability assessment is more demanding. But the pressures for more serious attention to sustainability concerns will inevitably deepen with the effects of continuing unsustainable practice. Moreover, as we have seen, environmental assessment practice is already evolving in the direction of sustainability assessment, and experimental applications are proliferating everywhere. It is therefore sensible to begin to consider not just the basic decision criteria and process design needs, but also some of the more difficult challenges of implementation.

TABLE 5.3 The basic design features of best-practice sustainability assessment processes

A best-practice sustainability assessment process

- begins with explicit commitment to sustainability objectives and to application of sustainability-based decision criteria and trade-off rules that give integrated attention to social, economic, cultural, political, and environmental factors and their interrelations
- incorporates means of specifying and integrating the general sustainability decision criteria and trade-off rules for the local and broader context of particular cases
- covers all potentially significant initiatives, at the strategic as well as project level, in a way that connects work at the two levels
- ensures that proponents of undertakings and responsible authorities are aware of their assessment obligations before they begin planning and that they have effective motivations (legal requirements or the equivalent) to meet these obligations

TABLE 5.3 The basic design features of best-practice sustainability assessment processes (cont'd)

A best-practice sustainability assessment process

- focuses attention on the most significant undertakings (at the strategic and project levels) and on work that will have the greatest beneficial influence
- is transparent and ensures open and effective involvement of local residents, potentially affected communities, and other parties with important knowledge and concerns to consider and an interest in ensuring properly rigorous assessment
- takes special steps to ensure representation of important interests and considerations not otherwise effectively included (e.g., disadvantaged populations, future generations, broader socio-ecological relations)
- is initiated at the outset of policy, program, and project deliberations when problems and/or opportunities are identified
- requires critical examination of purposes and comparative evaluation of alternatives in light of the sustainability-based decision criteria
- addresses positive as well as negative, indirect as well as direct, and cumulative as well as immediate effects
- recognizes uncertainties and requires estimates of confidence in effects predictions
- seeks to identify alternatives that offer the greatest overall benefits and that avoid undesirable trade-offs (rather than merely enhance/mitigate the effects of already chosen options)
- emphasizes enhancement of multiple, mutually reinforcing benefits as well as avoidance or mitigation of negative effects
- specifies and applies explicit trade-off rules, including requirements for explicit rationales for trade-off decisions
- favours options that reflect a precautionary approach to significant risks and incorporate adaptive design, and requires preparation for continuous learning and adaptive implementation
- is enshrined in law, with effective means of ensuring compliance with process requirements and decisions
- includes means of enforcing terms and conditions of approval, monitoring implementation and effects, and ensuring appropriate response to identified problems and opportunities through the full life cycle of assessed undertakings
- facilitates efficient implementation
- is integrated into a more complete framework that links strategic- and project-level assessment and places both as contributors to and beneficiaries of a larger regime for the pursuit of durable and desirable futures.

SOURCE: Adjusted slightly from Gibson et al. 2005, 146–7.

MAJOR IMPLEMENTATION CHALLENGES AND TASKS

In addition to basic decision criteria and general process design guidance, transition to sustainability assessment will require case-specific process guidance (appropriate processes for elaboration of the general process rules and the basic decision criteria for specific places and undertakings) and suitable methodologies (for sustainability deliberations as well as for baseline data, indicators, systems depictions, and desired future scenarios about and approaches to conflict resolution, for example concerning trade-offs).

Some of the necessary work is already underway in the broader realm of sustainability initiatives. Development of sustainability objectives and indicators, including locally and regionally specific ones, has been supported by many organizations and jurisdictions for more than a decade. Tools for integrating multiple lay stakeholders in evaluation and decision processes (through scenario-building, design charettes, valued ecosystem component identification, site selection criteria development, community mapping, etc.) are becoming increasingly well tested and sophisticated. Advanced methodologies for depicting complex systems and considering future changes in them are being applied at scales from the local to the global. As the already broad range of sustainability-oriented deliberations (urban planning, collaborative resource management, corporate greening, alternative national accounts, industrial ecology, growth management, etc.) continues to expand, it is reasonable to anticipate many further contributions of insight and methodology.

Sustainability assessments can also be expected to act as means of solving their own problems. Because they force more rigorous and better-integrated attention to sustainability requirements as the key concern of decision making in particular circumstances, they serve as a mechanism for clarifying general sustainability requirements, indicators, and decision rules, and for specifying them in particular contexts, through informed choices by the relevant parties. Nevertheless, it is worth looking a little more closely at some of the inevitably difficult issues. And given the present interest in approaches to effective integration, a good candidate for attention is the knotty problem of conflict and compromise.

The list of sustainability-based decision criteria in Table 5.2 ends with a requirement for integration. It demands that the first six requirements be pursued in mutually compatible ways that win positive effects all round and that precaution and adaptation be included in every case. There is no way around this. Significant and lasting improvements rely on linked, mutually supporting positive steps on all fronts. Perhaps this agreeable result can be achieved more often than we might expect. But existing examples are rare. In practice, compromises and trade-offs will be unavoidable in most program and project decisions, if only because overall global conditions are now so very far from sustainability.

ELABORATING APPROACHES TO TRADE-OFF DECISIONS

In narrower environmental assessments, trade-offs between biophysical or ecological considerations and competing social and economic objectives might be made outside the assessment framework. In sustainability assessment, all the policy commitments and all the development objectives are considered together and the trade-offs are addressed directly. The key trade-off issues can be summed up thus:

- Which objectives, potential damages, and promising gains are most significant (given that contribution to sustainability is the objective)?
- Which ones are most (or least) acceptable, in general and in particular circumstances?
- How should these decisions be made?

Common trade-off decisions include compensations and substitutions (direct and indirect compensation for, rather than full mitigation of, negative effects). For example, a gravel pit developer might offer later rehabilitation of a new operation on agricultural lands that are now at least somewhat degraded (a substitution in time). Or a subdivision developer might propose a constructed wetland to replace a relatively natural one (a substitution in place). Or a company exploring for oil and gas on traditional Aboriginal lands might offer a new community recreational facility to compensate for risks to traditional hunting or fishing (a substitution in kind). Also common are net gain and loss calculations (aggregation of net gain and net loss calculations). These might involve, for example, reduction of near-term ecological damage risks from surface storage of toxic wastes balanced against smaller but long-term risks from initially secure deep underground disposal (differences in time); or major damages to the interests of Aboriginal people displaced by a new dam balanced against more material security for larger numbers of poor farmers downstream (differences in place); or efficiency gains from industrial process improvements balanced against associated job losses (differences in kind, across requirements).

Even where sustainability objectives are widely understood and commonly accepted, different interests are likely to reach different conclusions about which compensations and net calculations may be justified. The answers will often also depend on the details. Just how serious are the losses, risks, and gains involved? Just how inequitable is the distribution of effects?

There are two interdependent approaches to guiding such trade-off decisions: focusing on the substantive issues and focusing on the processes for trade-off decision making.

Substance rules. Sustainability-based environmental assessment regimes can clarify application of the sustainability requirements by setting out general rules, or at least guidelines, for decisions about what sorts of trade-offs may or may not be acceptable. These can be complemented by more specific region- or sector-specific clarifications.

Nevertheless, perhaps few set rules will be appropriate for all cases (given different communities, cultures, ecosystems, stresses, aspirations, capacities, etc.), even within particular regions or sectors.

The one clearly essential general rule is that trade-off decisions must not compromise the fundamental objective of net sustainability gain. Also unacceptable is the sacrifice of a significant existing quality or a lasting major improvement for a trivial and transitory satisfaction. Some additional possibilities are suggested in Table 5.4. But often the acceptability of a trade-off will depend on the circumstances. It may be possible to agree on regional or local or even case-specific trade-off substance rules that take these circumstances into account. But how these particular rules are set becomes a matter of controversy. Here we enter the realm of process and the need for guidance on how the decisions are to be made.

TABLE 5.4 Possible general rules for decisions about trade-offs and compromises

Net gains

Any acceptable trade-off or set of trade-offs must deliver net progress towards meeting the requirements for sustainability; it must seek mutually reinforcing, cumulative, and lasting contributions and must favour achievement of the most positive feasible overall result, while avoiding significant adverse effects.

Burden of argument

Trade-off compromises that involve acceptance of adverse effects in sustainability-related areas are undesirable unless proven (or reasonably established) otherwise; the burden of justification falls on the proponent of the trade-off.

Avoidance of significant adverse effects

No trade-off that involves a significant adverse effect on any sustainability requirement area (e.g., any effect that might undermine the integrity of a viable socio-ecological system) can be justified unless the alternative is acceptance of an even more significant adverse effect.

- Generally, then, no compromise or trade-off is acceptable if it entails further decline or risk of decline in a major area of existing concern (e.g., as set out in official international, national, or other sustainability strategies or accords or as identified in open public processes at the local level), or if it endangers prospects for resolving problems properly identified as global, national and/or local priorities.
- Similarly, no trade-off is acceptable if it deepens problems in any requirement area (integrity, equity, etc.) where further decline in the existing situation may imperil the long-term viability of the whole, even if compensations of other kinds, or in other places are offered (e.g., if inequities are already deep, there may be no ecological rehabilitation or efficiency compensation for introduction of significantly greater inequities).
- No enhancement can be permitted as an acceptable trade-off against incomplete mitigation of significant adverse effects if stronger mitigation efforts are feasible.

Protection of the future

No displacement of a significant adverse effect from the present to the future can be justified unless the alternative is displacement of an even more significant negative effect from the present to the future.

TABLE 5.4 Possible general rules for decisions about trade-offs and compromises (cont'd)

Explicit justification

All trade-offs must be accompanied by an explicit justification based on openly identified, context-specific priorities as well as the sustainability decision criteria and the general trade-off rules.

- Justifications will be assisted by the presence of clarifying guides (sustainability policies, priority statements, plans based on analyses of existing stresses and desirable futures, guides to the evaluation of 'significance', etc.) that have been developed in processes as open and participative as those expected for sustainability assessments.

Open process

Proposed compromises and trade-offs must be addressed and justified through processes that include open and effective involvement of all stakeholders.

- Relevant stakeholders include those representing sustainability-relevant positions (e.g., community elders speaking for future generations) as well as those directly affected.
- While application of specialized expertise and technical tools can be very helpful, the decisions to be made are essentially and unavoidably value-laden and a public role is crucial.

SOURCE: From Gibson et al. 2005, chap. 6.

Process rules. Because the substance rules are insufficient, sustainability assessment regimes must provide guidance on selection and use of appropriate processes for making context-specific trade-off decisions. The processes for considering the options can include use of some of the many more or less elaborate tools (systems analysis, scenario-building, cost-benefit analysis, risk assessment, multi-stakeholder negotiation, etc.) that have been developed for formal decision making about trade-offs. But while expertise and technical tools can be very helpful, trade-off decisions are essentially and unavoidably value-laden. What and whose values are able to play a role in the design and application of tools, and in the use of deliberative processes, is therefore crucial.

Table 5.4 sets out a list of possible trade-off rules that include both substance and process components. This list, like the list of core sustainability criteria, is open to debate and in need of elaboration that is not possible here. But it illustrates the kind of guidance that should be provided in sustainability assessment regimes.

NEXT STEPS

Sustainability assessment has so far been explored mostly through particular initiatives undertaken in more or less special circumstances. The proliferation of such initiatives seems likely to continue, if only because there are so many real problems that demand attention to intertwined socio-economic/political and biophysical/ecological considerations and require a long-term perspective. Often this will involve creation of ad hoc processes. But sometimes it will be possible to make creative use of existing legislated regimes or to establish new mechanisms with sustainability assessment capacities.

Expanded versions of existing strategic- and project-level environmental assessment processes have great potential as vehicles for sustainability assessment. They have been evolving, albeit unevenly and not everywhere, in the direction of sustainability assessment, and most environmental assessment processes incorporate more of the basic design features of best-practice sustainability assessment processes today than they did 10 and 20 years ago.

Further progress in this direction is both plausible and desirable. It is not entirely risk free, however. One of the great challenges of environmental assessment processes has been to force attention to factors that have been generally neglected in conventional decision making. Effects on ecosystems and communities are now much more likely to be noted and taken seriously than they were in the years before environmental assessment. But the gains so far have been limited and remain fragile in many jurisdictions. Steps to introduce broader sustainability assessment should root environmental considerations more deeply in the core of deliberations and decisions at the strategic as well as project levels. But because sustainability assessment integrates the ecological and community concerns with other social, economic, and political factors, badly designed sustainability assessment processes could lead to less direct attention on environmental issues and reverse some of the hard-won gains of the past three decades.

New or adjusted assessment processes that ensure attention to the full suite of sustainability requirements and incorporate all of the basic process characteristics listed above are unlikely to threaten any past gains. But putting such processes in place is not likely to be achieved in one step. The risk lies in ill-conceived or poorly implemented incremental changes.

Two complementary solutions are available. The first is to continue efforts to clarify sustainability assessment aims and requirements. The better we understand the objective, the less likely we are to go astray in implementation efforts. The second is to accept the precautionary reliance on diversity. As noted above, experiments with sustainability assessment or its equivalent have been and are being undertaken not just in environmental assessment regimes but also in land-use planning, site restoration, corporate greening, community-level development assistance, trade option evaluation, and a host of other fields. Moreover, they are using not just conventional law and policy tools but also certification schemes, corporate behaviour codes, ethical investment criteria, sustainable livelihood analyses, multi-stakeholder collaborations, and a long list of other mechanisms. Errors and missteps in any one of these areas will be minimally dangerous so long as the same basic agenda is being pursued on many other fronts.

If these steps are taken, the benefits of sustainability assessment should be available without sacrifice of environmental gains. Indeed, if sustainability assessment gives us a more comprehensive, open, and powerful means of ensuring careful attention to, and effective integration of, environmental concerns in decision making, we will have much enhanced prospects of lasting benefits all round.

Practical initiatives to build environmental assessment regimes into vehicles for more fully integrated sustainability assessment are already well underway. Moving them along will entail work on a variety of fronts. Certainly it will be crucial to revise laws, processes, and guidance material to clarify sustainability-based criteria, specify trade-off rules, and facilitate practical transition from the mitigation focus of conventional assessment to positive contributions to overall sustainability improvement. But no less important will be a host of particular debates, experiments, and innovations where a mix of relevant interests, experts, and citizens put their heads together to solve problems or pursue opportunities. When these involve a shared commitment to multiple objectives, inclusive process, careful deliberation, and lasting solutions, they will in their own ways contribute to the advancement of sustainability assessment and the enhancement of our understandings about to how to pursue a more desirable and durable future.

ACKNOWLEDGMENTS

This chapter is based in part on 'Sustainability Assessment: Basic Components of a Practical Approach,' a paper presented at the International Association for Impact Assessment annual conference in Vancouver, British Columbia, 24–30 April 2004. It also draws from a recent book, *Sustainability Assessment: Criteria and Processes*, referenced below, and from earlier work on these matters for the Canadian Environmental Assessment Agency and the Canadian International Development Agency. Neither of the latter agencies is responsible for or certain to accept any of the positions presented here.

NOTES

1. This is the case with the Canadian federal approach to strategic-level assessment. See CEAA 2004.
2. These include approaches designed for the mining industry (MMSD 2002), urban regions (Ravetz 2000; Devuyst et al. 2001), infrastructure development (Arce and Gullón 2000), and trade liberalization agreements (Kirkpatrick and Morrissey 1999).
3. These include advances in the study of complex systems, especially in ecology and resource management (cf. Gunderson et al. 1995; Gunderson and Holling 2002), and other sociopolitical and biophysical realms. For example, the field of new governance recognizes the power and limitations of market mechanisms, has doubts about the potential adequacy of state interventions, challenges 'civilizing missions' and universal solutions, and accepts context dependency and expanded 'governance' roles for other tools and players (cf. Dryzek 1992; Paehlke 2003; Sachs 1999; Beck 1995).
4. For a discussion of the pillars approaches, see Mebratu 1998. The Canadian International Development Agency (CIDA 1997, chap. 2) has favoured a five-pillar approach.
5. The sets of sustainability criteria prepared for environmental assessment applications by Clive George (1999) centre on present and future equity, combining ecological and socio-economic considerations. Keith Pezzoli (1997) carried out a transdisciplinary review of sustainable development literature and identified the four key challenges as holism and

co-evolution, social justice and equity, empowerment and community building, and sustainable production and reproduction. Neil Harrison (2000, 99–118) found three key concentrations in the literature—efficiency, equity, and ethics—judged each of them too limited and mechanical, and proposed to incorporate them all within an emphasis on building social capacity for flexibly adaptive action. Non-pillar categories are also common in criteria lists developed for practical planning applications centred on 'quality of life issues'. See, for example, the United Kingdom's criteria for regional sustainable development work (UK 1999).

REFERENCES

André, P., C.E. Delisle, and J-P. Revéret. 2004. *Environmental assessment for sustainable development: Processes, actors and practice*. Trans. Brigitte Koelsch. Montreal: Presses Internationales Polytechnique.

Arce, R., and N. Gullón. 2000. The application of strategic environmental assessment to sustainability assessment of infrastructure development. *Environmental Impact Assessment Review* 20:393–402.

Aschemann, R. 2004. Local Agenda 21 of Graz City: The development of its indicators. Paper presented at the 2004 conference of the International Association for Impact Assessment, Vancouver, BC, 26 April.

Beanlands, G., and P. Duinker. 1983. *An ecological framework for environmental impact assessment in Canada*. Halifax: Institute for Resource and Environmental Studies, Dalhousie University.

Beck, U. 1995. *Ecological politics in an age of risk*. Cambridge: Polity Press.

Bekhuys, T.J., and G. McKay. 2004. Incorporating sustainability into environmental impact assessment: A case study of the Vancouver 2010 Olympic and Paralympic Winter Games. Paper presented at the 2004 conference of the International Association for Impact Assessment, Vancouver, BC, 28 April.

CEAA (Canadian Environmental Assessment Agency). 2004. *The Cabinet Directive on the Environmental Assessment of Policy, Plan and Program Proposals*. Ottawa/Gatineau: CEAA. http://www.ceaa.gc.ca/016/directive_e.htm.

CIDA (Canadian International Development Agency). 1997. *Our commitment to sustainable development*. Ottawa/Hull: CIDA.

CSA (Canadian Standards Association), Working Group of the EIA Technical Committee. 1999. *Preliminary draft standard: Environmental assessment*, Draft #14, 26 July. Toronto: CSA.

Devuyst, D., with L. Hens and W. De Lannoy (eds). 2001. *How green is the city: Sustainability assessment and the management of urban environments*. New York: Columbia University Press.

Dryzek, J.S. 1992. Ecology and discursive democracy: Beyond liberal capitalism and the administrative state. *Capitalism, Nature and Socialism* 3 (2): 18–42.

George, C. 1999. Testing for sustainable development through environmental assessment. *Environmental Impact Assessment Review* 19:175–200.

Gibson, R.B. 1993. Environmental assessment design: Lessons from the Canadian experience. *Environmental Professional* 15 (1): 12–24.

——— 2000. Favouring the higher test: Contribution of sustainability as the central criterion for reviews and decisions under the Canadian Environmental Assessment Act. *Journal of Environmental Law and Practice* 10 (1): 39–54.

——— 2002a. *Specification of sustainability-based environmental assessment decision criteria and implications for determining 'significance' in environmental assessment.* Ottawa/Gatineau: Canadian Environmental Assessment Agency. http://www.ceaa-acee.gc.ca/015/0002/0009/index_e.htm.

——— 2002b. From Wreck Cove to Voisey's Bay: The evolution of federal environmental assessment in Canada. *Impact Assessment and Project Appraisal* 20 (3): 151–9.

Gibson, R.B., with S. Hassan, S. Holtz, J. Tansey, and G. Whitelaw. 2005. *Sustainability assessment: Criteria and processes.* London: Earthscan

Gunderson, L.H., and C.S. Holling. 2002. *Panarchy: Understanding transformations in human and natural systems.* Washington, DC: Island Press.

Gunderson, L.H., C.S. Holling, and S.S. Light (eds). 1995. *Barriers and bridges to the renewal of ecosystems and institutions.* New York: Columbia University Press.

Harrison, N.E. 2000. *Constructing sustainable development.* New York: SUNY.

Hodge, R.A., and I. Thomson. 2004. From minimal damage to net contribution: Mining's seven questions to sustainability. Paper presented at the 2004 conference of the International Association for Impact Assessment, Vancouver, BC, 28 April.

Kirkpatrick, C., N. Lee, and O. Morrissey. 1999. *WTO New Round: Sustainability Impact Assessment Study.* Phase One Report, October. Manchester: Institute for Development Policy and Management.

Lawrence, D. 2003. *Environmental impact assessment: Practical solutions to recurrent problems.* Hoboken, NJ: John Wiley and Sons.

Mebratu, D. 1998. Sustainability and sustainable development: Historical and conceptual review. *Environmental Impact Assessment Review* 18:493–520.

MMSD (Mining, Minerals and Sustainable Development Project). 2002. *Breaking new ground: Mining, minerals, and sustainable development.* London: International Institute for Environment and Development and World Business Council for Sustainable Development. http://www.iied.org/mmsd/finalreport.

Nelson, P., C. Azare, E. Sampong, B. Yeboah, and E. Darko-Mensah. 2004. SEA and the Ghana Poverty Reduction Strategy. Paper presented at the 2004 conference of the International Association for Impact Assessment, Vancouver, BC, 28 April.

Paehlke, R. 2003. *Democracy's dilemma: Environment, social equity and the global economy.* Cambridge, MA: MIT Press.

Pezzoli, K. 1997. Sustainable development: A transdisciplinary overview of the literature. *Journal of Environmental Planning and Management* 40 (5): 549–74.

Pope, J. 2004. Conceptualising sustainability assessment: Three models and a case study. Paper presented at the 2004 conference of the International Association for Impact Assessment, Vancouver, BC, 26 April.

Ravetz, J. 2000. Integrated assessment for sustainability appraisal in cities and regions. *Environmental Impact Assessment Review* 20:31–64.

Sachs, W. 1999. *Planet dialectics: Explorations in environment and development.* London: Zed Books.

Sadler, B. 1996. *Environmental assessment in a changing world: Evaluating practice to improve performance.* International Study on the Effectiveness of Environmental Assessment, Final Report, Canadian Environmental Assessment Agency and the International Association for Impact Assessment. http://www.ceaa.gc.ca/017/012/iaia8_e.pdf.

Senécal, P., B. Sadler, B. Goldsmith, K. Brown, and S. Conover. 1999. Principles of environmental impact assessment best practice. International Association for Impact Assessment and Institute of Environmental Assessment. http://www.iaia.org/Non_Members/Pubs_Ref_Material/pubs_ref_material_index.htm.

UK (United Kingdom). 1999. *A better quality of life, summary.* London: Government of the United Kingdom. http://www.sustainable-development.gov.uk/publications/uk-strategy99/index.htm.

Wood, C. 2003. *Environmental impact assessment: A comparative review.* 2nd edn. Harlow, Essex: Pearson Education.

Chapter 6

Integrated, Adaptive Watershed Management

Bruce P. Hooper and Chris Lant

INTRODUCTION

This chapter will explore the thesis that the dynamic nature of the challenges of water resources is driving a change in the intellectual foundations of management towards integrated, adaptive watershed management. Institutions for watershed governance are also changing in a manner that facilitates this form of management.

Concerns for water resource development, structural flood control, and centralized treatment of drinking water and waste water remain of great relevance in developing countries, where flooding claims many lives and an estimated 1.7 billion people lack access to safe drinking water, leading to 5 million deaths annually (WRI 2000). Fortunately, these issues have been substantially resolved in affluent countries. There the focus is shifting towards managing land-use patterns to prevent polluted runoff and groundwater contamination, restoring the physical integrity of rivers to reverse declines in aquatic ecosystems, and maintaining and improving the provision of ecosystem services. Riverine ecosystems are in decline owing to excessive water withdrawals, channel modifications, erosion and sedimentation, deterioration of substrate quality, chemical contaminants, overfishing, and introduction of exotic species (Adler et al. 1993; Doppelt et al. 1993; Karr and Schlosser 1978). Sediment, pathogens, and nutrients derived from runoff lead the list of pollutants for which Total Maximum Daily Load (TMDL) plans are required under Clean Water Act 303(d) in the United States. Similar concerns for water-quality management in deindustrializing countries have driven support for a regional approach to management in Western Europe (Weale et al. 2000). These profound ecological problems often arise as unintended consequences of management actions taken to meet, sometimes to optimize, subsystem goals (e.g., farm income, GDP) in the absence of appropriate informational and behaviour-modifying feedback on the effects of such subsystem behaviour on the larger system in which it is embedded. As Hawken et al. (1999) have observed, 'optimizing components in isolation tends to pessimize the whole system' (p. 117). To the extent that this is true, progress lies in better integration. But how is integration achieved? Here are some possible answers:

1. Integration is achieved by making *endogenous* what had been considered to be *exogenous*—that is, by expanding the definition of the system being managed.
2. Integration is achieved by incorporating an understanding of the *spatial* and *temporal* dynamics of the system being managed.
3. Integration is achieved by fostering *self-regulatory* and *self-organizing* feedbacks among system components.

These theoretical insights, derived from systems theory and its recent rejuvenation as complexity theory, are central to an understanding of integration and therefore of integrated watershed management. But for a rural land or water manager—a national government resource management officer, a hobby farmer, an operator of a large irrigation farm, a rancher, a commercial timber company executive, a state agency director, a producer of organic produce for the local farmers' market, a feedlot operator, a subsistence farmer in a developing country, the inheritor of a 100-hectare plot of land that has been in the family for generations—the problems of land and water management are overwhelmingly pragmatic, and nothing is more pragmatic than economic or political survival, unless it is getting that diesel or small gasoline engine to start. Rural land and water managers by and large are not interested in integrated management beyond their parcel, and even less in complexity theory. They want to solve the practical problems often associated with changes in technology, markets, or regulations. How can managers apply new biotechnologies while containing input costs and gaining consumer confidence in GMOs (genetically modified organisms)? How can small-field efficient irrigation techniques be implemented when there is not enough food to eat? How can no-till be made to work without the need to increase herbicide applications? How can managers make sure that their organic produce qualifies for the higher prices certification brings while still getting enough nitrogen to the crop to ensure good yields? What should be done with the huge manure pile at the feedlot that is now regulated as a point source by the US Clean Water Act? How can the Endangered Species Act be enforced in the face of gun-toting ranchers and demanding irrigators? How can steers be marketed in the middle of a mad cow disease scare? Is there a timber market for those 80-year-old oaks on the back half? These are the real issues that local land resource managers face. But the solutions they implement can constitute a tyranny of small decisions that governs larger ecosystems and watersheds by default.

Integrated approaches to watershed management have emerged in many countries over the last three decades in response to these overwhelmingly local needs, as well as to regional environmental concerns (Douglass 1972; Bowonder and Ramana 1985; Wang et al. 1989; Mitchell and Hollick 1993; Crockett 1995; Ahluwalia 1997; Born and Genskow 1999; Michaels 2001). Coordinated watershed management is not new. The notion of coordinating different aspects of natural resource management emerged in developments in several countries in the earlier part of the twentieth century. In the 1930s, New Deal conservation programs in the United States stimulated widespread interest in and adoption of soil and water conservation programs and basin-wide programs of employment creation, soil erosion mitigation, power production, river navigation improvement, poverty reduction, and reforestation such as those of the Tennessee Valley Authority, creating the era characterized by a multi-purpose approach to the development and management of river basins and the reservoirs that had been constructed in them (Krutilla and Eckstein 1958). Integrated watershed management today, however, is fundamentally different from this earlier approach in that the focus is on cross-sectoral coordination in the planning and management of natural resources throughout the watershed. The rise of the environmental movement of the 1960s and the years following stimulated the emergence of the integrated approach. In developed countries, the integrated approach is becoming more accepted philosophically and the desired outcomes are often clear (clean water, sustained water supplies, restored rivers, reduced erosion rates), but there is confusion and uncertainty about process.

PANARCHY IN OUR IDEAS

Mid-twentieth-century concepts and theories are being found increasingly inadequate to meet the needs of integrated solutions to ongoing economic and social dilemmas or to escalating environmental problems. Gunderson and Holling (2002) have made a bold attempt at integrating ecological, economic, and social theories in the context of long-term dynamics in their concept of *panarchy*. Systems evolve in a long-term cycle where they, like a roller coaster being dragged by a chain to the top of the track, slowly accumulate 'capital', conceived as order, information, or low entropy. Natural capital accumulates in the ecological form of biomass and tightly efficient nutrient cycles. Social capital accumulates as complex social hierarchies, power structures, and legal systems. Economic capital accumulates as well-developed infrastructures and productive labour systems for the employment of proven technologies. Like the potential energy contained in the roller-coaster cars as they near the top of the track, the capital-rich system becomes vulnerable to sudden release as exogenous shocks push it past a threshold. The reorganization phase of the ensuing low-capital system is a moment of both crisis and opportunity, with unfilled ecological and market niches and power vacuums that are opportunistically filled, setting the stage for the next capital-accumulating ecological, economic, or political regime. The reader will readily think of fire-, flood-, or pest-dominated ecosystems, Kondratieff waves and globalization, or Chinese dynasties and Western revolution as examples of regimes that pass through these stages of panarchy.

Thomas Kuhn's 1970 classic *The Structure of Scientific Revolutions* presents a similar dynamic with respect to our understanding of the natural and human worlds in the form of scientific theories or paradigms. 'Normal science' methodically answers puzzle-like scientific questions embedded in an explicit or implicit disciplinary paradigm. Successful paradigms are capable of explaining new phenomena as they are discovered as applications and routine extensions of existing theory. In this manner, intellectual capital steadily accumulates. But when problems arise that cannot be managed or anomalies arise that cannot be explained in the context of the dominant paradigm, a theoretical crisis emerges. New theories compete for disciplinary prominence and may ultimately replace the old paradigm, which is subsequently viewed as failed, obsolete, or incomplete. Scientific knowledge specific to the old paradigm must be reinterpreted or discarded. Like R-strategists and clever entrepreneurs, the crisis is an opportunity for scientists and scholars who can build prime niches of intellectual capital associated with the paradigm shift, so much so that many more paradigm shifts are claimed (especially at tenure review time) than in retrospect can be showed to have occurred.

Despite this caveat, the empirical challenges of seemingly intractable environmental problems, combined with the theoretical insights being derived from complexity theory, place resources management at a juncture in the panarchy cycle of our ideas where older prediction-, optimization-, and equilibrium-oriented paradigms derived from neo-classical economics, civil engineering, and Clementsian ecology are accelerating down the roller-coaster track of intellectual capital release. Scientific knowledge relevant to environmental management is being rapidly reorganized along new interdisciplinary paradigms founded in principles of complex systems interdependence, evolutionary change, and continuous adaptation. Yet theory construction within this new paradigm is just beginning as the management of pragmatic problems goes on more urgently than ever in pursuit of environmental sustainability. In short, we suggest that integrated resource management, the new paradigm for watershed management, needs further conceptual development and theoretical refinement.

CONCEPTUAL FOUNDATIONS OF INTEGRATED WATERSHED MANAGEMENT

A watershed can be usefully viewed as a storehouse of natural capital in the form of soil, flora and fauna, and so on that helps produce multiple marketable commodities, such as crops, livestock, and timber, and non-marketable ecosystem services, such as nutrient cycling, carbon sequestration, soil formation and binding, sediment trapping, and wildlife habitat (Gottfried 1992; WRI 2000). Empirical research in ecological economics clearly shows that greater investments in natural capital are needed in order to ensure delivery of ecosystem services in the present and the future (Bjorklund et al. 1999; Pretty et al. 2000). This is particularly true in the case of agriculture, which utilizes about half of global usable land and a similar proportion of its water and therefore has a preponderant effect on natural capital and delivery of ecosystem services while simultaneously having to meet the challenge of meeting food demands for a growing human population (Tilman et al. 2002). Unfortunately, there is a narrow range of social circumstances under which resource managers are willing to make substantial personal investments in the present to achieve even more substantial public benefits in the future. One critical implication is that ecosystem services, and the natural capital that generates them, will, under capitalist regimes, be under-produced because they are public goods and are viewed by landowners as positive externalities.

Figure 6.1 provides a conceptualization of watersheds as spatially defined, complex human-environment systems. Social driving forces in particular watersheds reflect macro-scale social structures and forces (e.g., globalization, property rights to

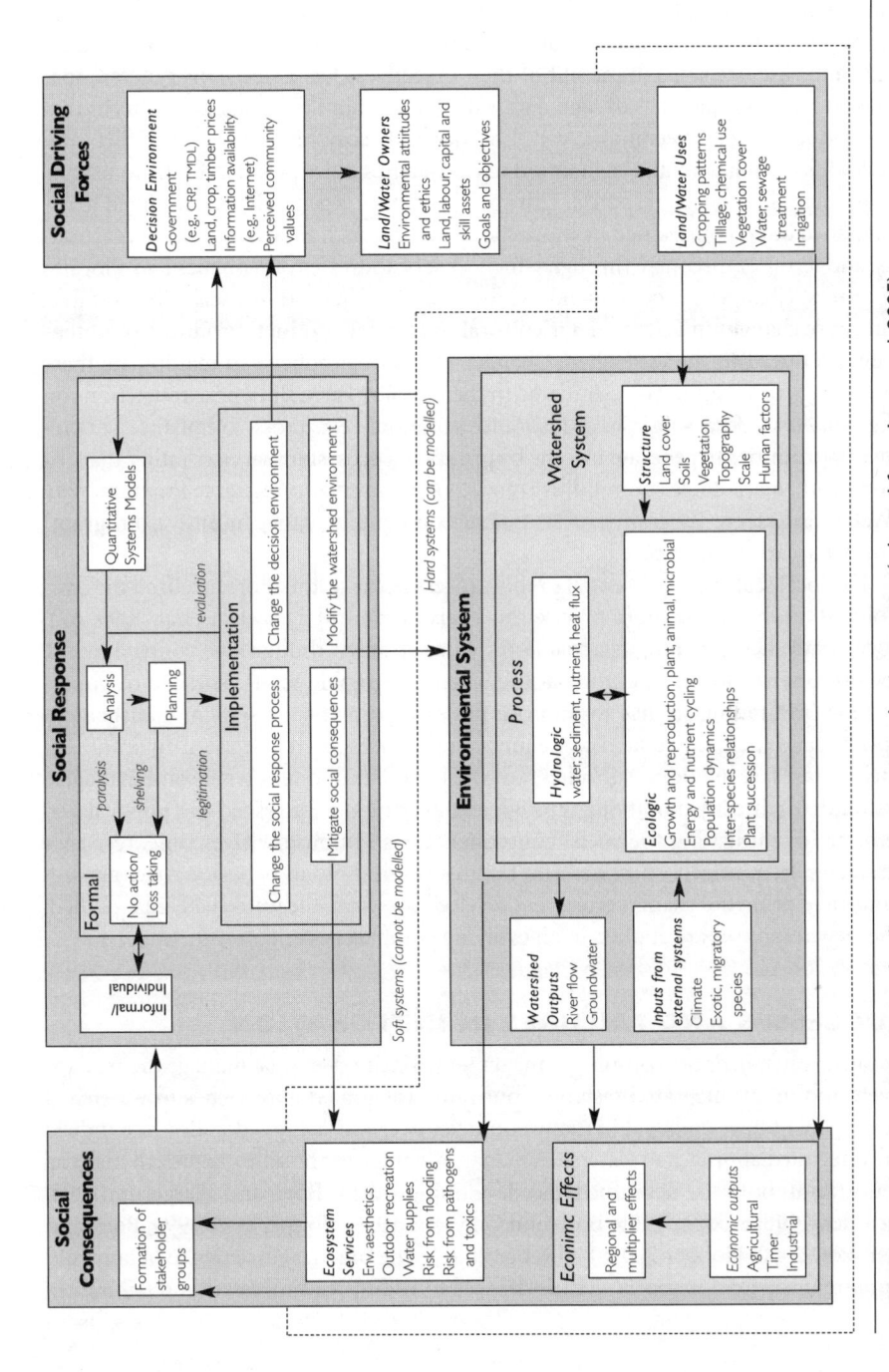

FIGURE 6.1 A watershed as a complex human-environment adaptive system (adapted from Lant et al. 2005)

resources, demographic trends, technological capacities, intellectual orientations and values), as they are manifested and reproduced differentially as variable decision environments from locality to locality. As agents, land and water managers respond to the opportunities and constraints presented by these social structures by making land- and water-use decisions in pursuit of their objectives. These decisions generate spatial and temporal patterns of land and water use within the watershed, patterns that co-produce a suite of economic commodities and ecosystem services. Conflicts and trade-offs among potential economic and ecosystem service products of the watershed set the stage for a social response. This social response is generally not confined to the watershed but is constrained by the social structures such as laws, governmental bodies, and terms of discourse through which society addresses other issues and allocates resources. Changes in socio-economic forces at the international scale, for example, can cause changes in prices for agricultural or forestry products produced or potentially produced in the watershed. Changes in national policy can manipulate these prices (e.g., price supports re-initiated in the US 2002 Farm Bill), pose restrictions on eligibility for federal subsidies or insurance programs (e.g., cross-compliance), or create opportunities to generate income by producing ecosystem services rather than, or along with, marketable commodities (e.g., CRP [Conservation Reserve Program, WRP [Wetlands Reserve Program], or EQIP [Environmental Quality Incentives Program] contracts, carbon farming).

The local scale is also essential to this process because 'the more localized the control and feedback, the more precise the levels of control' (Hawken et al. 1999, 67). Local watershed planning is a viable response strategy, though one constrained by socio-economic forces operating at the larger geographic scales. Possible responses include (a) funding science to improve understanding of the system, management options, and their anticipated consequences; (b) changing the decision environment (prices, taxes, regulations, subsidies) to which land and water owners and managers respond; (c) directly modifying through engineering the watershed as a physical system; or (d) mitigating the social consequences as symptoms. Over time, response strategies are pursued, successes and failures result, human behaviours are altered, values and priorities change, ecosystem service flows from the watershed are modified, and new responses are applied in an evolving complex system.

SUCCESSES AND FAILURES IN INTEGRATION

Because watersheds are complex management systems, we must manage them adaptively and in an integrated fashion—but how? Integrated approaches to watershed management have emerged in many countries over the last two decades, and indeed the integrated approach to watershed management is the prevailing approach in many countries in both the developed and developing world (Born and Margerum 1993; Dearden 1996; Hooper 1995; Born and Genskow 1999; Hooper et al. 1999; Born and Genskow 2000; Hooper 2002; Ballweber 2002). The practical experience of land and water management in recent decades has led to widespread successes and failures. In summary, Hooper et al. (1999) identified 10 constraints to implementation. These constraints focus on a range of institutional, organizational, and human resource management issues and form the fundamental constraints to the integration approach in watershed management:

1. *Establishing the need for integration.* Integration is seen as a 'good thing' to do, but integration may bring net costs owing to the need for institutional reforms that lead to increased staff costs, higher costs to resource managers, and costs that may be inequitably distributed in time and space. Incurring these costs must be justified by the benefits of integration in solving discrete problems associated with linkages among systems. Otherwise, an individual or a single-agency response is all that is needed.

2. *Scope, ambiguity, and vagueness.* The integrated approach may be defined too broadly in a watershed management planning process, and this leads to superficial understandings of natural resource management. A focused approach is needed that identifies key drivers in a system and monitors these drivers in the form of indicators of system states. Management then focuses on manipulating these key drivers.

3. *Dubious premises.* An integrated approach often assumes that cooperation and consensus among interests are achievable and that the outcomes achieved by local voluntary interest groups will be the same, or better than, those achieved by line agencies in natural resource management. Another assumption is that voluntary groups should always be only advisory and should share responsibility with line agencies. In real problem-solving situations, either of these assumptions could be false.

4. *Importance of context.* This refers to how some aspects of a resource and environmental management decision are specific to a time and place. Many resource management problems may take decades to solve, so solutions should be at least intergenerational—say 25 years. But not all resource management problems are so time dependent, and local, immediate solutions can be made. This means that context should always drive management methods.

5. *Visibility, identity, leadership, and communication.* There is often confusion between the roles of integrated watershed management, economic development plans, and environmental management. Success comes from local/regional leadership where communication between stakeholders about the plans and actions of these three activities occurs at that scale.

6. *Coordination and collaboration.* Resistance to power sharing among and within government agencies is difficult to overcome. There is the need to share power and coordinate vertical and horizontal power sharing, yet some problems can easily marginalize a power group or are zero-sum games: painless coordination is severely limited.

7. *Information—availability, accessibility, and integration.* Lack of integration of information and lack of access to integrated information curtail effective cross-jurisdictional decision making in watershed management. Institutional

arrangements should be put in place to facilitate data and information integration and access, such as incentives for data agencies to share such information.
8. *Connections with local and regional planning.* Local planning traditionally applies to urban areas, but most watersheds are rural. Traditional regional planning tends to be dominated by economic growth, industrialization, and urban development. As a result, integrated watershed management is often marginalized by entrenched development processes. Watershed management should be incorporated into local and regional planning processes.
9. *Financial arrangements, market processes, and property rights.* These are often dismally ignored in watershed management. Who finances watershed management? Ongoing financial commitment and rolling budgets to maintain watershed management are needed. These arrangements must address questions of private property rights and who pays for common-pool resource management disbenefits. These issues are yet to be resolved in many watershed management programs.
10. *Conflict resolution.* Many resource management problems in a watershed are intractable, 'messy', and 'wicked' (without obvious solution). They touch different stakeholders in varying degrees in time and space, and a process for reducing conflict is frequently missing from the management system. The resolution to judicial approaches is often the default solution to resolving conflict.

CHARACTERIZING ADAPTIVE MANAGEMENT IN A WATERSHED

An adaptive approach provides a way in which complex resource management problems can be addressed by emphasizing management flexibility (Walters 1986). This goes well beyond traditional multi-purpose planning or sustained-yield approaches. It attempts to address economic development and resource use from an ecological perspective and to identify the sustainable limits of resource use. When critical thresholds are passed, ecosystem health indicators decline, perhaps irreversibly, indicating that land- and water-use practices are not sustainable. Human history provides many examples of societies that declined precipitously as a result of failures in soil conservation or watershed management (Rice and Rice 1984; Diamond 2004). In a forestry operation, on a farm, in an urban development proposal, in an industrial operation, or in a nature conservation program, the approach is similar: define coordinated management actions, implement those actions, audit outcomes, describe lessons learned, and revise management accordingly. These lessons learned appear to emerge to best effect when the core decision makers in a watershed management process come together to discuss management options. Perhaps adaptive watershed management was first put in practice in efforts to restore salmon runs in the Pacific Northwest of the United States, when the Northwest Power Planning Council adopted an adaptive policy to fish and wildlife management (Lee and Lawrence 1986; Volkman and McConnaha 1993) in which cooperation between stakeholders was deemed a function of human values and high levels of regional efficacy and autonomy (McGinnis 1995).

Grayson et al. (1994) applied Walters's adaptive environmental assessment and management approach to the development of water-quality management plans in Victoria, Australia (Walters 1986), using a computer simulation technique. They concluded that it was not the modelling procedure that was of primary importance, but rather the modelling workshops. They found that such workshops were 'a highly efficient medium for accumulation of information about the system and require participants to focus clearly on problems and achievable solutions from the outset.' The need for human interaction and adaptability was also found in work focusing on local watershed management modelling in Illinois (Seitz 1989) and in salinity management research in the Liverpool Plains, Australia (Hooper 1995).

Kai Lee's 1993 classic on adaptive management, *Compass and Gyroscope*, provides our best account to date of the political and ecological experience of integrated watershed management through adaptive management. Lee (1993) defines adaptive management and adaptive policies—originally developed by Canadian ecologist C.S. Holling in the 1970s at the International Institute of Applied Systems Analysis in Vienna—in a number of ways:

> Adaptive management is an approach to natural resource policy that embodies a simple imperative: policies are experiments; learn from them. (p. 9)

> Adaptive management is treating economic uses of nature as experiments so that we may learn efficiently from experience. (p. 8)

> Adaptive policies define experiments probing the behavior of the natural system. Experiments often bring surprises, but if resource management is recognized to be inherently uncertain, the surprises become opportunities to learn, rather than failures to predict. (p. 56)

> An adaptive policy is one that is designed from the outset to test clearly formulated hypotheses about the behavior of an ecosystem being changed by human use. (p. 53)

The last quote raises political and ethical issues in the implementation of experimental policies. Gilbert White, on the one hand, has long advocated post-audits, a form of civil science, as a means to maximize what social learning is possible from the examination of the consequences of past policies. Lee (1993) takes this further, however, to include hypothesis-testing power in policy design *a priori*, not *post-hoc*. As with the treatment of a medical patient, the scope of experimental treatments that are ethical expands as the condition worsens. This may be a fair analogy for restoring wild salmon runs to the upper Columbia River, the context of Lee's study. In a broader sense, Lee emphasizes that we do not know whether sustainability is achievable or how to achieve it, and we have no working examples of sustainability in industrial societies, whose markets link the products of local ecosystems to global economic demand. The conclusion appears to be that passive adaptive management, the White approach of scientifically evaluating the consequences of policies that have been and are being implemented for non-scientific purposes, is characteristic of all intelligent management and is essential for integrated management. Active adaptive management, Lee style, has thus far achieved few demonstrable successes in social learning over and above the passive, less politically risky, form. In order to justify its political risks, we should ask not only 'Do we know what the ailment is?' but also 'Is the patient that sick?' Thus, as ecological conditions worsen, adaptive management, like medical intervention, necessarily takes on a more and more active form.

Adaptive management, then, has two congruent, interdependent objectives: resource use and ecosystem sustainability. These require the use of an interdisciplinary approach, one that depends for its success on lateral thinking, learning from previous experience, inter-agency cooperation, strong leadership, and extensive involvement by stakeholders at all levels of watershed decision making. This approach, then, develops (a) institutional, (b) organizational, and (c) human resource management improvements as the key tools to implement the integrated, adaptive approach (Table 6.1).

It is in this context that adaptive management can be starkly contrasted with planning. Effective planning requires that two conditions hold: first, that the planner has jurisdictional authority over the system being planned, and second, that the planner can predict the consequences of his or her plans. It is clear that, especially in multiple ownership watersheds (which is most of them), neither of these conditions holds. 'Integrated Watershed Planning' is therefore not the title of this chapter. Adaptive management, in contrast to planning, is a process based on continuous feedback from management actions and a theory-guided interpretation of the measurements of the evolving system and its interacting components that together serve to minimize the error in our trials resulting from complexity and unpredictability. This is accomplished by treating the anticipated effect of management actions on the system as hypotheses and rigorously testing these hypotheses by monitoring the system response. This informational feedback comes in the form of data, which the manager interprets as indicators that build 'stories' based on theoretical knowledge and prior experience with the system (Kranz et al. 2004). Clearly, adaptive management calls for long-term commitment achieved through mechanisms of institutional capacity and memory. In reading the following definition, readers may think they are reading about raising small children, or about a doctor treating a patient with a chronic syndrome, rather than about adaptive management:

> The overall goal of adaptive management is not to maintain an optimal condition of the resource, but to develop an optimal management capacity. This is accomplished by maintaining ecological resilience that allows the system to react to inevitable stresses, and generating flexibility in institutions and stakeholders that allow managers to react when conditions change. The result is that, rather than managing for a single, optimal state, we manage within a range of acceptable outcomes while avoiding catastrophes and irreversible negative effects. (Johnson 1999, 8)

TABLE 6.1 Institutional, organizational, and human resource elements of integrated, adaptive watershed management

Characteristic	Description
Coordinated rather than amalgamated programs of action	The approach emphasizes bringing together initiatives from relevant government agencies, community groups, and private sector organizations and identifies coordinating actions between programs, functional overlaps, and missing actions; a stratecic plan of action (perhaps expressed as a watershed management plan) prescribes team-based management plans that require stakeholders to work together for specific programs.
Top-down meeting; bottom-up management	The approach recognizes the value of both top-down (from government) and bottom-up (from local actions by community groups and local agencies) working together to achieve mutually agreed upon objectives.
Strategic planning rather than all-embracing efforts	In any large watershed, there is always the challenge to deal with the tyranny of small decisions and address them all at once. The preferred approach favours selecting those issues that are of paramount importance politically as well as critical to the health of regional ecosystems and the vitality of regional economies.
A regional perspective	The approach recognizes the need for a coordinating mechanism, frequently at the sub-state or inter-state level across large watersheds and/or river basins. The scale of resource management planning and management is meso-scale—at scales of 1:100,000 to 1:500,000. The approach brings together bioregional-scale resource management with basin-scale surface hydrology and provincial-scale groundwater management.
Adaptive rather than linear approaches to resource management planning	The watershed management plan documents learning experiences in implementing coordinated actions; it is more likely to be a loose-leaf folder than a bound book in that it allows review and refinement of coordinated actions; and once these actions are implemented and reviewed for effectiveness,

TABLE 6.1 Institutional, organizational, and human resource elements of integrated, adaptive watershed management (cont'd)

Characteristic	Description
	lessons are learned and ongoing actions are modified to improve outcomes.
Holistic rather than single or multi-purpose management	The approach favours the management of several resource and environmental problems simultaneously and recognizes that individual resource-use systems (mines, industries, forestry operations, farms, fisheries) operate within a larger regional context and must be managed within the ecological thresholds (limits) of those systems; otherwise resource degradation will occur. The critical task is to identify both the thresholds beyond which irreversible change will occur in ecosystems and social systems and the acceptable limits of change in those systems.
Reactive resource use planning	The approach uses modelling to predict the outcomes of current and potential practices, tests the possible outcomes against real experiences, and reports these experiences back to the watershed stakeholder community. These 'experiments' in natural resource management are documented in a watershed management plan to serve as a learning archive for future watershed managers.
Creative/cost-effective rather than prescriptive financial management, using cost sharing	Development of cost-sharing processes for different activities is seen as a way to leverage program funds, increase community ownership of watershed management activities, and increase efficiency of integrated resource management investments; and involves creatively mixing funds from other government programs, industry groups, and the watershed community in an attempt to leverage national funds.
Adaptive management styles	Cooperative rather than confrontational management. Committed rather than command-and-control management. Empowered rather than directed staff. Problem-solving rather than functional staff.

TABLE 6.1 Institutional, organizational, and human resource elements of integrated, adaptive watershed management (cont'd)

Characteristic	Description
Flexible rather than rigid organization structures	Learning from previous experiences will frequently require organizations to change their structures and functions to provide improved outcomes. The approach requires a supple organizational structure, one that can easily adapt to new contingencies.

SOURCE: Adapted from Hooper 1995.

In practice, integrated, adaptive watershed management can mean different watershed management actions in different contexts. It has been successful in Chesapeake Bay of the United States (Hennessey 1994). But does it work elsewhere? What works best in one watershed may not work best in another. The same is true at the basin or valley scale. For example, Shah et al. (2003), who have worked at the river basin scale, suggest that a different approach is needed in developing countries, one that recognizes fundamental differences in climatic and ecosystem behaviours, social organization, and institutional capacity. They suggest that issues such as famine, economic development through irrigation, poverty reduction, and provision of potable water need to be addressed as priorities in developing countries; these issues are considered less critical in developed countries, as they have been addressed at earlier periods of economic development in river basins.

Integrated, adaptive watershed management processes can potentially be introduced into local government planning, but this depends on whether local governments have the legal jurisdiction to implement legislation relating to natural resource planning. One option might be a watershed management act that would define prescriptive roles and regulations for local government (Ruhl et al. 2003). To this could be added the determination of a range of organizational options for improved watershed governance, ranging from loose coalitions of local initiatives (similar to current practices in the United States) to catchment management authorities (such as in Australia) or integrated river basin management authorities such as the Mekong River Basin Commission.

At the strategic level, national policies for resource management frequently fail to link economic development and ecosystem management. Ecologically sustainable development (ESD) principles often give only limited recognition to resource management. In Australia, this weakness was seen as partly due to poor communication and collaboration between national government agencies (AACM 1995). The integrated, adaptive watershed management process can be an effective tool for the implementation of ESD initiatives, but there is limited understanding of how ESD approaches could be implemented at the river basin, subcatchment, or bioregional level. Similarly, the development of natural resource management policy needs to be linked to other state and nationally based ESD initiatives, such as programs for tourism, national energy, and the preservation of rare and endangered species.

At the national policy level, more work needs to be done on the use of economic instruments in integrated, adaptive watershed management. Economic tools like cost-benefit analysis need to be used to quantify benefits and disbenefits of management options, and other tools, such as multi-criteria analysis, could be used to include other non-monetary values of resource management options and evaluate them alongside monetary values (Prato 1998). There is scope for using market mechanisms as a means to determine ecosystem benefits and demonstrate how the market can be used to encourage spending on management practices that enhance environmental values. This approach requires a clear definition of property rights in a watershed situation in order to untie the ownership of resources from the property they have been assigned (such as water rights) and allow the resource to be traded to improve efficient use. A recent innovation in Australia, for example, has demonstrated that there is political support for this approach, yet it remains to be tested (Council of Australian Government 2004). Much work remains to be done in the developing and testing of alternative arrangements for the various problems requiring action. However, enough is known to allow researchers to establish some examples and get on with testing the theory. For example, water reallocation among urban, industrial, agricultural, and ecological uses is highly amenable to market mechanisms. Recreational water resource users need to be able to bid for water rights (e.g., in-stream flow for anglers and boat owners or wetland flow for hunters) along with industrial, agricultural, and urban users. The cost of water should include the full cost of extraction, treatment, reticulation, waste water return, effluent treatment, and water return. Water utilities and river management authorities provide vehicles for implementing market mechanisms, so long as they have clearly established criteria for charging for water supply and effluent management, managing water quality, monitoring water quantity and quality, and internalizing all external costs associated with the extraction, supply, and treatment of water and waste water (AACM 1995). Ultimately, policy needs to grapple with the proposition that questions of sustainability—or, more precisely, of the great unknown involved in claims on ecosystems—call for the creation of rights that intentionally restrict the operation of commodity production and markets in order that ecosystem services be maintained.

BUILDING MODELS TO FACILITATE INTEGRATIVE, ADAPTIVE WATERSHED MANAGEMENT

How should we structure models of watersheds that serve the purpose of integrated, adaptive management by capturing watersheds' complexity, their adaptive mechanisms, and their evolutionary nature, and that can test hypotheses and answer research questions while avoiding the pitfalls of optimization, prediction, and equilibrium thinking inherent in many current models? A model's structure should reflect the questions that we need it to answer. Critical researchable questions for integrated, adaptive watershed management might include the following:

1. How do land and water managers respond to ecological, economic, social, and policy stimuli and spatial and temporal feedbacks through resource decision making?
2. How do these responses produce landscape patterns at watershed scales?
3. How do landscape patterns influence watershed ecosystem function?
4. How do spatial and temporal variations in watershed ecosystem function influence the provision of ecosystem services and availability of natural resources?
5. What are the possible combinations of marketable commodities and ecosystem services that can be produced by a watershed, given ecological, economic, technologic, and other constraints?
6. What is the nature and shape of the ecological-economic production possibilities frontier (EEPPF) so formed (see Figure 6.2)?

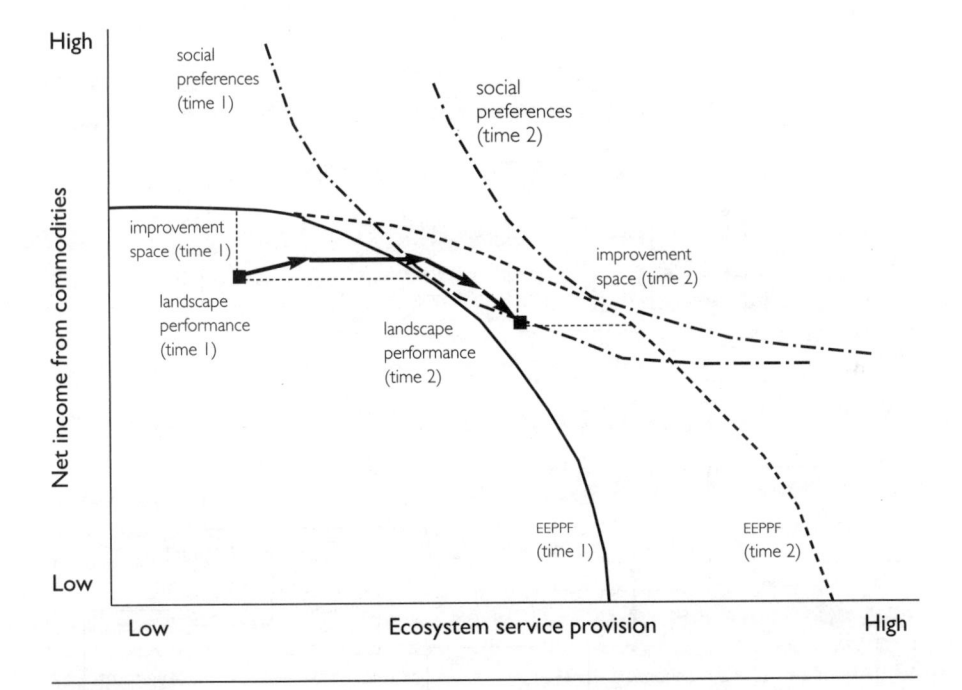

FIGURE 6.2
An ecological-economic production possibilities frontier (EEPPF) and social preferences indifference curves are shown for time period 1 (the present) and time period 2 (some point in the future). The landscape that maximizes total social utility at any one point in time is found at the point where these curves are tangent. Present landscape performance is shown as inside the EEPPF with the space between current performance and the EEPPF as the improvement space. Through adaptive management, landscape performance moves over time through the improvement space towards the social utility maximizing point on the EEPPF. Both social preferences and the EEPPF, however, change over time and are thus moving targets. In this example, ecosystem service provision is expanded over time, but this need not be the case.

7. What is the performance of the current landscape with respect to this multi-dimensional EEPPF?
8. How can managers and policy-makers identify actions that will improve watershed performance towards or along the EEPPF with respect to selected objectives?
9. Can manipulation of individual policy variables improve the ecological-economic performance of watersheds?
10. How does adaptive management function as a feedback mechanism between landscape performance relative to societal objectives and management actions designed to achieve those objectives?
11. What type of organizations are best suited to implement integrated, adaptive watershed management, and how can their performance be measured?

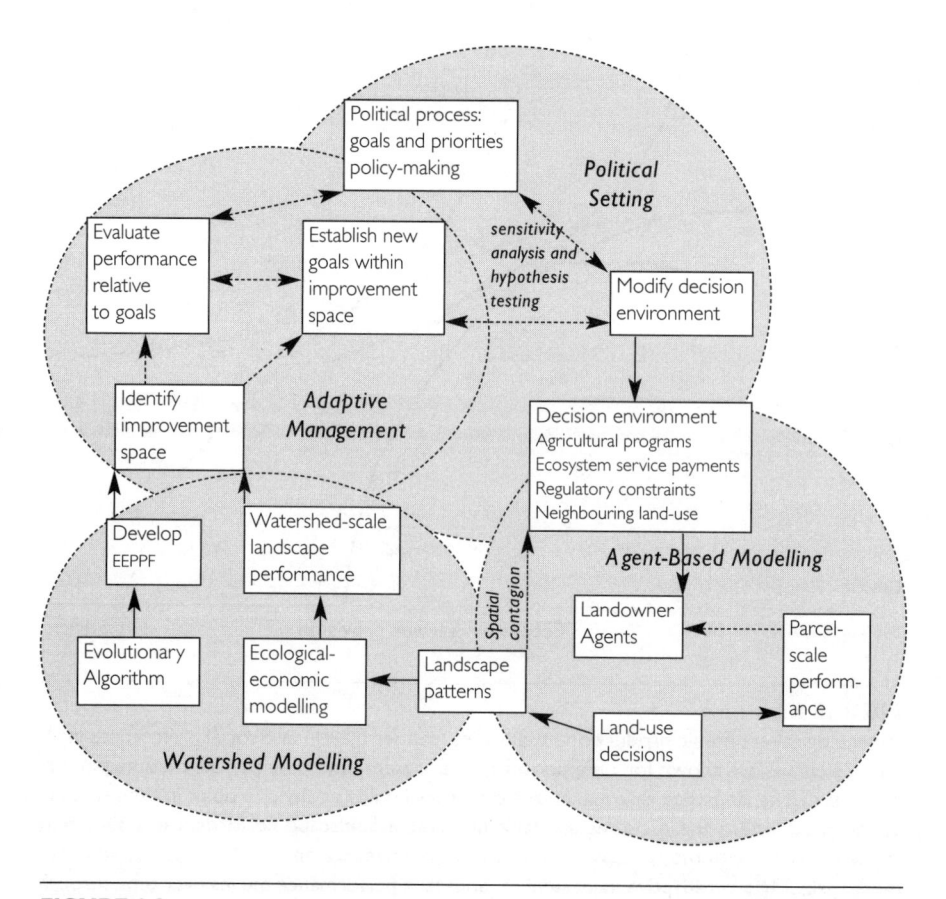

FIGURE 6.3
An approach to watershed modelling to facilitate integrated, adaptive management within a political context

12. What are the relative time scales at which, on the one hand, management actions designed to improve provision of specific commodities and ecosystem services can become effective and, on the other, social preferences for these commodities and ecosystem services change? That is, can adaptive management get society what it wants from watersheds before society decides it wants something else?

The US National Research Council study *New Strategies for America's Watersheds* (1999) emphasizes the need for new models that utilize GIS (geographic information system) and specifically incorporate an understanding of human responsiveness to policy and other factors as evaluative tools in managing watersheds. Figure 6.3 represents one possible blueprint for a watershed modelling system that achieves this in the context of integrated, adaptive management in a political context. This blueprint consists of two largely independent modelling approaches whose results are then compared in an adaptive management process. The first approach identifies the EEPPF for the watershed as a whole. Points on or within the EEPPF are associated with

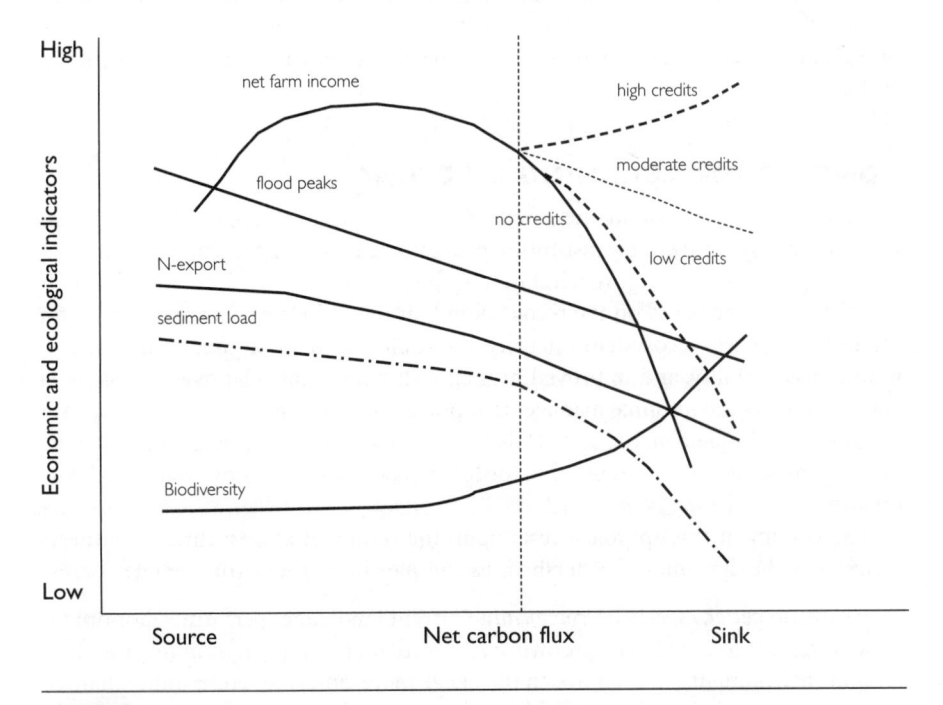

FIGURE 6.4
The hypothesized qualitative relationship between net carbon flux and other economic and ecosystem service goals. Net sequestration is positively associated with all other ecosystem service goals. Net farm income without carbon credits is maximized at a slight net carbon source. Carbon credits of varying level could maximize farm income at a state of net carbon sink.

specific land-use patterns. The axes of the PPF are economic products and ecosystem services provided by the watershed. The EEPPF will evolve on a periodic basis as a result of technological change, climate disturbances, and evolving landscapes (Figure 6.2). While economic theory assumes that PPFs are continuous and convex, implying competitive trade-offs among economic and ecological goals, PPFs can be concave when goods and services are complementary. For example, it is hypothesized that carbon sequestration is complementary with biodiversity, flood peak reduction, sediment and nitrogen export, and farm income if carbon credits are set at a high enough level (Figure 6.4). Under these circumstances, output sets resulting from alternative resource allocations display complex behaviour (Brock et al. 2002). This complex behaviour can perhaps be understood through the use of evolutionary algorithms.

Discovery of the watershed-scale EEPPF establishes what the watershed can theoretically provide but ignores how land and water managers respond to stimuli in making decisions for their parcel. The second modelling approach would therefore use agent-based modelling to predict the resulting landscape pattern and performance associated with sets of policy variables contained in the decision environment. As discussed by Parker et al. (2003), this approach incorporates essential feedbacks into the modelling process, thereby capturing the dynamics of non-linear behaviour evident in real landscapes. Policy variables can then be manipulated as the basis of scenarios with performance evaluated by the agent-based model with respect to the EEPPF.

CONCLUSION: LEARNING BY DOING

In this chapter we propose an integrated, adaptive management approach to overcome what appear to be intractable institutional, organizational, and human resource management problems. This approach should be built on an improved, rigorous theoretical construct, one based on the recognition that natural resources are managed for multiple purposes—ecosystem integrity, the resilience of social organizations, economic sustainability, and improved human well-being. The adaptive management approach offers the resource management practitioner a suite of tools to be used for diagnosis and implementation, tools that can characterize the resource management decision-making setting and provide coordinated actions to improve watershed governance.

The efficacy of this approach rests upon the results of at least three hypotheses concerning the dynamics of watersheds as complex human-environmental systems:

1. *We can do better, especially ecologically.* Current landscapes perform suboptimally with respect to the EEPPF, but owing to reward systems inherent in historic decision environments, they approach the EEPPF more closely for commodity than for ecosystem service production. If false, the status quo is the best that can be done, or alternatively, economic development, not integrated management with an ecological focus, is needed.
2. *Space matters.* Due to economies of configuration, the ecological-economic performance of a watershed can be improved through watershed-level management

over and above parcel-level management of land and water use. In response to changes in policy variables, land-use changes occur in a non-linear and patchy manner due to neighbourhood effects. If false, watershed-scale integration and a concern with thresholds are not needed.

And perhaps most importantly, but also most subtly:

3. *We're running on a winding road, not a treadmill.* Adaptive management can improve watershed-scale performance for specified ecological and economic objectives at a time scale that is relevant with respect to changing social preferences, political structures, and broad environmental changes. If false, by the time we make progress towards our ecological and economic objectives, society has changed the objectives or the environment has changed so fundamentally that our efforts are irrelevant.

ACKNOWLEDGMENTS

The authors thank Bruce Mitchell, our country, Canada, and anonymous reviewers for their comments on an earlier version of this chapter.

Figure 6.1 is reprinted from *Ecological Economics*, 55 (4), Lant, C.L. Kraft, S.E., J. Beaulieu, D. Bennett, T. Loftus, and J. Nicklow, Using GIS-based ecological-economic modeling to evaluate policies affecting agricultural watersheds, 467–84. Copyright (2005). With permission from Elsevier, whom the authors thank.

REFERENCES

AACM International and Centre for Water Policy Research. 1995. *Enhancing the effectiveness of catchment management planning.* Final report for the Department of Primary Industries and Energy. Adelaide: AACM International Pty Ltd.

Adler, R.W., J.C. Landman, and D.M. Cameron. 1993. *The Clean Water Act: 20 years later.* Washington, DC: Island Press.

Ahluwalia, M. 1997. Representing communities—The case of a community-based watershed management project in Rajasthan, India. *IDS Bulletin* 28 (4): 23.

Ballweber, A. J. 2002. Prospects for comprehensive integrated watershed management under existing law. *Water Resources Update*, no. 100 (Summer): 19–27.

Bjorklund, J., K.E. Limburg, and T. Rydberg. 1999. Impact of production intensity on the ability of the agricultural landscape to generate ecosystem services: An example from Sweden. *Ecological Economics* 29:269–91.

Born, S.M., and K.D. Genskow. 1999. *Exploring the 'Watershed approach'—critical dimensions of state-local partnerships: The Four Corners watershed innovators initiative final report.* Madison, WI: Department of Urban and Regional Planning, University of Wisconsin-Madison.

———. 2000. *Towards understanding new watershed initiatives: A report from the Madison Watershed Workshop.* Madison, WI: Cooperative Extension Service, University of Wisconsin-Madison.

Born, S.M., and R. Margerum. 1993. *Integrated environmental management: Improving the practice in Wisconsin*. Madison, WI: Department of Urban and Regional Planning at the University of Wisconsin-Madison.

Bowonder B., and K. Ramana. 1985. Management of watersheds and water resources planning. *Water International* 10 (3): 121–31.

Brock, W.A., K.G. Maler, and C. Perrings. 2002. Resilience and sustainability: The economic analysis of nonlinear dynamic systems. In L.H. Gunderson and C.S. Holling (eds), *Panarchy: Understanding transformations in human and natural systems*, 261–89. Covelo, CA: Island Press.

Council of Australian Governments. National Water Initiative. In Communiqué, Council of Australian Governments' Meeting, 25 June 2004. http://www.coag.gov.au/meetings/250604/index.htm. Accessed 6 July 2004.

Crockett, S. 1995. *The watershed management approach to resource protection*. Proceedings of the annual Underground Injection Control and Ground Water Protection Forum. Oklahoma City: Ground Water Protection Council.

Dearden, P. 1996. Integrated watershed management planning and information requirements in northern Thailand. *Canadian Journal of Developmental Studies* 17 (1): 31–51.

Diamond, J. 2004. John Chaffee Lecture on science and the environment, January 2004, Washington, DC.

Doppelt, M., M. Scurlock, C. Frissell, and J. Karr. 1993. *Entering the watershed: A new approach to save America's river ecosystems*. Washington, DC: Island Press.

Douglass, J.E. 1972. *Annotated bibliography of publications on watershed management*. SE-93. Washington, DC: USDA, Forest Service.

Gottfried, R.R.1992. The value of a watershed as a series of linked multiproduct assets. *Ecological Economics* 5 (2): 145–61.

Grayson, R.B., J.M. Doolan, and T. Blake. 1994. Application of AEAM (adaptive environmental assessment and management) to water quality in the Latrobe River Catchment. *Journal of Environmental Management* 41:245–58.

Gunderson, L.H., and C.S. Holling (eds). 2002. *Panarchy: Understanding transformations in human and natural systems*. Covelo, CA: Island Press.

Hawken, P., A. Lovins, and L.H. Lovins. 1999. *Natural capitalism: Creating the next industrial revolution*. New York: Little, Brown and Company.

Hennessey, T.M. 1994. Governance and adaptive management for estuarine ecosystems: The case of the Chesapeake Bay. *Coastal Management* 22 (2): 119–45.

Hooper, B.P. 1995. Adoption of best management practices for dryland salinity: The need for an integrated environmental management approach. Results of a study in the Goran Catchment, NSW. Centre for Water Policy Research, University of New England, Armidale.

——— 2002. Towards more effective integrated watershed management in Australia: Results of a national survey, and implications for urban catchment management. *Water Resources Update* 100:28–35.

Hooper, B.P., G.T. McDonald, and B. Mitchell. 1999. Facilitating integrated resource and environmental management: Australian and Canadian perspectives. *Journal of Environmental Planning and Management* 42 (5): 746–66.

Johnson, B.L. 1999. The role of adaptive management as an operational approach for resource management agencies. *Conservation Ecology* 3 (2): 8. http://www.consecol/org/vol3/iss2/art1.

Karr, J.R., and I.J. Schlosser. 1978. Water resources and the land-water interface. *Science* 201:229–34.

Kranz, R., S.P. Gasteyer, H.T. Heintz, R. Shafer, and A.D. Steinman. 2004. Conceptual foundations for the sustainable water resources roundtable. *Water Resources Update* 127:11–19.

Krutilla, J.V., and O. Eckstein. 1958. *Multiple purpose river development*. Baltimore, MD: Johns Hopkins Press.

Kuhn, T.S. 1970. *The structure of scientific revolutions*. 2nd edn. Chicago: University of Chicago Press.

Lant, C.L., S.E. Kraft, J. Beaulieu, D. Bennett, T. Loftus, and J. Nicklow. 2005. Using GIS-based ecological-economic modeling to evaluate policies affecting agricultural watersheds. *Ecological Economics* 55 (4): 467–84.

Lee, K.N. 1993. *Compass and gyroscope: Integrating science and politics for the environment*. Washington, DC: Island Press.

Lee, K.N., and J. Lawrence. 1986. Adaptive management: Learning from the Columbia River Basin Fish and Wildlife Program. Paper presented at Symposium on Salmon Law: Restoration under the Northwest Power Act, Northwest School of Law, Lewis and Clark College. *Environmental Law* 16:431.

McGinnis, M.V. 1995. On the verge of collapse: The Columbia River system, wild salmon and the Northwest Power Planning Council. *Natural Resources Journal* 35 (1): 63–92.

Michaels, S. 2001. Making collaborative watershed management work: The confluence of state and regional initiatives. *Environmental Management* 27 (1): 27–35.

Mitchell, B., and M. Hollick. 1993. Integrated catchment management in Western Australia: Transition from concept to implementation. *Environmental Management* 17 (6): 737–43.

Mitchell, B., and D. Shrubsole. 1994. *Canadian water management: Visions for sustainability*. Cambridge, ON: Canadian Water Resources Association.

National Research Council, 1999. *New strategies for America's watersheds*. Washington, DC: National Academy Press.

Parker, D.C., S.M. Manson, M.A. Janssen, M.J. Hoffmann, and P. Deadman. 2003. Multi-agent systems for the simulation of land-use and land-cover change: A review. *Annals, Association of American Geographers* 93:314–37.

Prato, T. 1998. *Natural resource and environmental economics*. Ames, IA: Iowa State University Press.

Pretty, J.N., C. Brett, D. Gee, R.E. Hine, C.F. Mason, J.I.L. Morison, H. Raven, M.D. Rayment, and G. Van der Blij. 2000. An assessment of the total external costs of UK agriculture. *Agricultural Systems* 65:1134–6.

Rice, D.S., and P.M. Rice. 1984. Lessons from the Maya. *Latin American Research Review* 19 (3): 7–34.

Ruhl, J.B., C. Lant, T. Loftus, S. Kraft, J. Adams, and L. Duram. 2003. Proposal for a model state watershed management act. *Environmental Law* 33 (4): 929–48.

Seitz, W. 1989. University of Illinois-Extension, personal communication.

Shah, T., D. Molden, and R. Sakthivadivel. 2003. Limits to leapfrogging: Issues in transposing successful river basin institutions in the developing world. Water policy briefing, IWMI-TATA Water Policy Program, Colombo, Sri Lanka.

Tilman, D., K.G. Cassman, P.A. Matson, R. Naylor, and S. Polasky. 2002. Agricultural sustainability and intensive production practices. *Nature* 418:671–7.

Volkman, J.M., and W.E. McConnaha. 1993. Through a glass darkly: Columbia River salmon, the Endangered Species Act, and adaptive management. *Environmental Law* 23:1249.

Walters, C. 1986. *Adaptive management of renewable resources.* New York: Macmillan.

Wang, L., X. Hong, and B. Xie. 1989. The planning method of small watershed management. In Ding-Lianzhen (ed.), *Proceedings of the Fourth International Symposium on River Sedimentation.* China: China Ocean Press.

Weale, A., G. Pridham, M. Cini, D. Konstadakopulos, and B. Flynn. 2000. *Environmental governance in Europe. An ever closer ecological union?* Guildford, Surrey: Oxford University Press.

Woolridge, M., and N.R. Jennings. 1995. Intelligent agents: Theory and practice. *Knowledge Engineering Review* 10:115–52.

World Resources Institute. 2000. *World resources 2000–2001: People and ecosystems: The fraying web of life.* Washington, DC: World Resources Institute.

CHAPTER 7

Implementation in a Complex Setting: Integrated Environmental Planning in the Fraser River Estuary

Kevin S. Hanna

The complex nature of natural resource and environmental issues and the confines imposed by administrative jurisdictions and physical boundaries have often relegated models of integration to intra-jurisdictional or intra-agency application. But the interconnectedness of environmental issues makes integration in management and planning attractive and necessary. Such approaches offer the potential to address what Slocombe and Hanna (in this book) call a persistent problem in environmental management—the fragmentation of authority across agencies and jurisdictional levels. But just as fragmentation can be difficult to overcome, even a modest approach to integration can prove challenging to implement.

This chapter provides a framework for characterizing hindrances and challenges to implementing integrated approaches to resource and environmental management. It presents an example of implementation in a complex setting—a case that illustrates one practicable approach to overcoming obstacles. The case study[1] presented here examines how one approach to integrated resource management, the Fraser River Estuary Management Program (FREMP) in Vancouver, British Columbia, was influenced at a very early stage by an understanding that implementation of any change in the administrative status quo would be difficult. Each participant in the FREMP process recognized the importance of developing an integrated approach that could be applied successfully within an environmentally and jurisdictionally complex locale.[2] Since many resource and environmental issues are set within intricate and conflict-prone settings, the FREMP case may have a particular resonance with the experiences of those working in other places.

The Fraser River estuary provides an interesting example of resource management for several reasons. The estuary area includes the city of Vancouver and has a regional population of about two million people (Figure 7.1). Over the last three decades, the region has seen one of the highest urban growth rates in North America, but despite such pressure it still contains large areas of rural land and relatively untouched river and marine shorelines (Figure 7.2). The environmental resources of the estuary are considerable. The Fraser is one of the world's great salmon-producing systems, and its estuary contains riparian habitat important for the inward and outward migration of Pacific salmon. Although the Fraser may be in good shape relative to other large

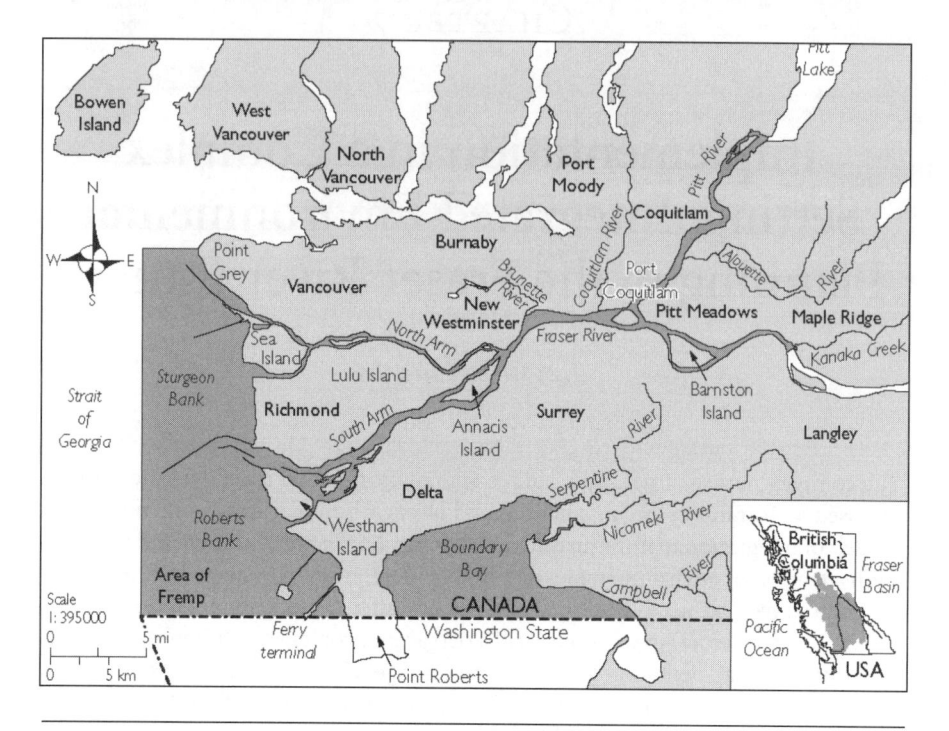

120 | INTEGRATED RESOURCE AND ENVIRONMENTAL MANAGEMENT

FIGURE 7.1 Area of the Fraser River Estuary Management Program

FIGURE 7.2 Fraser River estuary shoreline near Point Grey

FIGURE 7.3 Log booms at Iona Island

global rivers, it often seems on the threshold of unwelcome change. Its estuary hosts a range of industrial activities, many of which affect the river (Figure 7.3).

In the estuary region, responsibility for management and planning policy rests with three levels of government—federal, provincial, and regional/municipal. In the past this set the stage for conflict about how to manage the river's resources, but this changed with the formation in the 1980s of a low-key cooperative framework, the Fraser River Estuary Management Program. Ideally, an integrated approach not only addresses the activities of a broad range of agency actors, but is also inclusive, welcoming government and non-government stakeholders alike. The FREMP, however, concentrates on agencies and in particular on a few key departments. Participation by those outside government is ad hoc and can hardly be described as pluralistic. The program's impacts are indirect and achieved by facilitating actions—especially coordinating planning policy and environmental reviews—rather than by capital projects or new regulations. This highlights a challenge in the evaluation of some policies and programs. Explicit cause/effect relationships between policy and physical or social outcomes are not always easily apparent where capital works or new regulations are not the dominant instruments. Success can be difficult to measure in settings where the links between program actions and environmental change may be indirect.

A FRAMEWORK FOR CHARACTERIZING IMPLEMENTATION CHALLENGES

At one time there was an assumption that once a decision was made, its execution became a simple and mundane affair that did not merit significant attention (Hyder 1984). When it came to program or policy efficacy, it was the quality of the idea that mattered and not so much its execution. Some agencies still approach the policy process under this assumption. Of course, the evaluation literature has matured, and while ideas matter, when it comes to putting them into practice, even the best can go awry. Effective implementation is integral to the success of policy ideas.

Implementation can largely be defined in two ways: first, as the stage between decision and action (Hessing and Howlett 1997) where the underlying issue is whether or not the decision will, or can, actually be realized in a manner consistent with the policy objective (Brekke 1987); and second, as a continuing activity where the focus is on analyzing the way that policies or programs function (Brekke 1987; Freeman 1980; Rossi and Freeman 1985). The latter implies the monitoring, auditing, and refining of ongoing activities. Implementation is a dynamic process, one that often involves negotiation, compromise, and shifting goals (Ham and Hill 1993). It is not a purely administrative event that can be evaluated in terms of which components do or do not perform as expected or required; rather implementation is a policy/action continuum in which an interactive and negotiative process occurs over time (Ham and Hill 1993). Negotiative elements emerge especially when there are adversarial aspects to a policy process. Since integration often requires a change in agency relationships—and even the most modest integration efforts will mean some change in power relationships—adversarial dynamics can be an important consideration when integrated approaches are conceived and applied.

Overlapping themes have been common in discussions of the challenges to policy implementation. These have been variously described as 'activities common to the (implementation) process' (Jones 1984), 'obstacles to be overcome' (Mitchell 2002), 'conditions to be met for implementation success' (Sabatier and Mazmanian 1981), essential factors that determine policy success (Van Meter and Van Horn 1975), or factors that shape instrument choice (Linder and Peters 1989). Whatever the title or rubric they are gathered under, common elements (or macro-obstacles) can be assembled into a framework that has particular relevance to integrated approaches. Seven key obstacle realms have emerged in the implementation literature (cf. Baker 1989; Linder and Peters 1989; Margerum 1999; Margerum and Born 1995, 2000; Mitchell 2002; Sabatier and Mazmanian 1981; Walther 1987; and Weale 1992):

- *Tractability.* While tractability is an essential issue in policy implementation, it runs the risk of being used to avoid difficult issues. Tractability implies complexity of the context. It includes factors such as the physical boundary, the extent to which power relationships would have to be restructured, the extent of stakeholder behaviour change required, and the degree to which the problem is solvable by the policy instrument. It also suggests resource intensiveness, including not only

administrative costs but also operational simplicity. In essence, given the program design and the dynamics of the problem setting, how manageable is the problem?
- *Clarity of objectives.* Clear objectives not only support the articulation of an implementation strategy, but also can be used to delineate the responsibilities of stakeholders. Clarity asks that practitioners and enabling frameworks (such as legislation, regulations, or agreements) be structured to maximize the probability that target groups understand their obligations and will perform as desired (Sabatier and Mazmanian 1981). Clear objectives will also aid evaluation of the program's success (Mitchell 2002). But there may be a tendency to avoid clarity in order to get a program 'off the ground'. Integration programs in complex settings may begin with vague objectives in order to encourage participation by and the support of diverse stakeholders. But a deliberate vagueness is also favoured by adaptive approaches—it allows for flexibility as the social/economic or environmental dynamics of the setting change. There may be an assumption that expectations will be more clearly articulated as a program develops, as more information about the extent of a problem is collected, or as more resources are allocated.
- *Knowledge of cause and effect relationships.* If policy is to be successful, there must be an understanding of the links between activities and goals (Mitchell 2002). However, beyond involving knowledge of the impact of program activities, understanding cause-and-effect relationships means knowing how biophysical and other systems impact each other. Implementation requires ensuring that the enabling framework incorporates sound theory, identifying the primary factors and causal links that affect policy objectives, and also identifying the target groups whose activities affect the problem and, ultimately, program success.
- *Commitment.* Stakeholder commitment to program success is an essential aspect of implementation (Mitchell 2002). But not all stakeholders are created equal. Commitment by key agencies, resource managers, or the political level often determines success, despite the rhetoric of public consultation and broad stakeholder support that often accompanies integration. In some respects, this can be characterized in terms of gaining legitimacy among key players (Margerum and Born 1995; Margerum 1999; Walther 1987; Mitchell 1986). With legitimacy will likely come meaningful commitment. The challenge of commitment also means ensuring that the policy/program is supported by the constituency groups who hold authority, and that sufficient jurisdiction is granted to those responsible for implementation. Commitment is demonstrated not only through authority but also through the allocation of sufficient resources.
- *Information.* Data gaps and uncertainty about the nature of the problem/issue limit the ability of actors to identify causal linkages or the likelihood that the enabling framework will achieve the policy objective(s). Information also presents a paradox. On the one hand, data may be an essential aspect of a program's development, aiding understanding of the temporal and spatial aspects of the issues to be managed. But on the other, there are consistent challenges in collecting, processing, and incorporating scientific knowledge into policy processes (cf. Susskind

et al. in this book). Ballard and Fortmann (also in this book) suggest that integrative actions might be more likely to involve interactions about which there is little scientific information. Information development can be a key activity of an integrated approach before specific actions can be decided upon. Information development has its own challenges. Data collection, or 'studying the problem', might also be seen by some as a stalling tactic. Thus, perpetual research can become a way of avoiding doing something about the problem.

- *History.* Walther (1987) writes that successful implementation and the impacts of IRM (integrated resource management) are largely a function of the historical context into which the program is inserted. The argument is that legacies of conflict and cooperation, planning, resources, and timing (the stage of the problem at which it is implemented) are dominant elements in implementation success. Walther recognizes other elements, but the historical context, notably with respect to conflict and cooperation legacies, can impose significant implementation challenges.
- *Power of personalities.* Integration requires, indeed demands, interaction. Agency, business, and conservation interests are presented by individuals, who bring their biases, ideologies, personal experiences, and likes and dislikes to a process. In resource and environmental management, conflict, often acrimonious, is common, and having to overcome the well-entrenched and frequently ardent positions can limit the potential and efficacy of an integrated approach. In some settings, successful implementation may depend very much on who is at the table and whether they get along with, listen to, and like each other.
- *Unique challenges.* Since each policy setting provides unique challenges, factors that may be unique to the setting, policy, or program should also be considered. Mitchell (2002) refers to cultural challenges, notably within the context of the developing world. Rayner (1991) notes the importance of ideological perspective in multi-jurisdictional implementation, and of varied approaches to managerialism, based in political economic ideology. Corruption, political influence, and nepotism may emerge as issues quite apart from all others. But unique challenges can emerge beyond cultural nuances—for example, the personalities of individuals may be especially conducive, or not, to implementing an integrated approach; the influence or hegemony of an industry may limit integrated efforts; or the attributes of a community in terms of its identifying with a resource industry could negate a willingness to change management approaches or diversify resource use. There can also be a disconnect between the policy instruments available to implementing agencies and the expectations or culture of affected communities (cf. Wilmsen in this book). Or perhaps a form of trap exists, where the need for change is recognized, but social or economic factors render stakeholders seemingly unable to implement it—this can be akin to Platt's (1973) characterization of a *social trap*.

The characterization of implementation is not conducive to the application of a single template. With respect to integrated approaches, the implementation process is often evolutionary, sometimes ad hoc or adaptive, and is commonly affected by inconsistencies in budgets, statutory authority, political imperatives, the nuances of public interest, and communication and information—frequent issues in public policy processes. Two common criticisms of environmental agencies centre on their role as supporters or detractors of policy and the ineffectiveness of bureaucratic structures in addressing environmental problems (Dryzek 1990; Paehlke 1990; Torgerson 1990a and 1990b). These characteristics are reflected in the fragmented nature of the resource and environmental management processes. Responsibilities for economic, social, and environmental management are divided among agencies, or levels of government. Agencies also have established constituencies whose ideologies influence the ability or willingness of administrative strictures to implement policy, especially when such may be new or controversial or counter to the interests of a dominant constituency. Nor are bureaucracies always uniform entities in terms of organization, sense of purpose, or internal cohesion. A supportive bureaucracy is essential to implementation, but in some instances bureaucrats will be responsible for putting into effect policies that they have had little influence in formulating.

Despite a broad acceptance of the need for integrated approaches in resource and environmental management, progress in implementation has been hesitant and unsystematic. In part this hesitancy results from the obstacles outlined above, but it also reflects a reliance on processes where participants learn as they go with few examples and no universal model to follow or learn from (Mitchell 1990). This has often led to cautious processes where participants follow incremental strategies (Mitchell 1990). A problem associated with many integrated strategies may be, not that the models are unworkable, but that the scale of attention often lacks tractability. Mitchell (1986) suggests that many models have been unrealistically broad in scope, whereas a well-defined approach to integration maintains an awareness of the linkages and interrelationships but keeps the process within manageable boundaries, both physically and in terms of issues (strategic in implementation and attention).

There are also instances where integration is introduced as a 'problem solving strategy into already complex situations that are characterized by predominant resource use conflict' (Walther 1987, 439). In this situation, integration might begin as a reactive measure and be thus more likely to succeed as part of a proactive approach.

North American environmental and resource planning can still be characterized in terms of closed policy networks (Skogstad and Kopas 1992, 47). Bureaucracies can work together and with key non-government interests to develop and implement policy. In many settings, industry has an integral, perhaps unduly influential, role in policy formulation. The relationship between government and industry has been described as client based, where the interests of capitalist stakeholders dominate processes (Coleman 1988; Skogstad and Kopas 1992). A key component of integration is the improvement of communication among stakeholders (Walther 1987, 440). While resource agencies might include a framework for inter-agency communication within their mission statement, joint action is rare and can indeed conflict with bureaucratic interests.

Perhaps the greatest challenge in implementing an integrated approach is not in convincing decision makers that integration is needed, but in achieving a better understanding of how to implement such approaches (Hanna 1999). From a critical perspective, this requires a change in power relationships that may constitute a significant challenge to tractability. Overcoming this challenge requires the construction of venues for deliberative practice, but without unnecessarily dismantling authority structures. The case that follows outlines the importance attached to implementation in the design of an approach to integrated planning.

THE FRASER RIVER ESTUARY MANAGEMENT PROGRAM

In Canada, jurisdiction over environmental and resource issues is fragmented among three levels of government—national, provincial, and local. Environmental responsibilities are far from precise, and jurisdictions overlap. Jurisdiction with respect to water resource authority can be particularly nebulous. By one account, responsibility for the Fraser River is divided among some 60 government agencies (McPhee and Wiebe 1986).

The 1970s saw rapid urban growth and rising public concern about water quality and fish and wildlife habitat in the estuary region. In 1977 the federal and provincial governments initiated the Fraser River Estuary Study (FRES). The FRES was a two-phase process. The first phase (1977 to 1978) described the state of the estuarine environment and outlined management options (plans) for balancing human uses and the estuary's ecological systems. This phase was guided by a steering committee composed of senior federal and provincial public servants. In 1978 the steering committee issued a report that called for the development of an 'estuary management plan' through a collaborative, integrated, inter-agency process. The committee understood that, given the inter-agency tensions that had been a part of estuarine management up to that time, a new level of formal governance, an estuary authority, would face resistance at both levels of government—even if it had been accepted and promoted at the political level. Recognizing that an estuary authority would be difficult to implement, the steering committee suggested the creation of an 'estuary council'. This body would have been a political entity with ultimate authority for establishing and implementing policy. Unlike the authority model, which would have been a new stand-alone agency, the council would have been a cooperative, with no unique authority, served by inter-agency groups that would have had the responsibility of providing information to, and implementing the policies articulated by, the council.

In the second phase (1979 to 1985), a multi-agency review committee was created to examine program options and prepare a strategy for implementing the recommendations made during the first phase. The second phase considered three possible approaches to management and planning. Each contained integration elements:

1. *A committee approach*, which would have sought voluntary consensus on an advisory management program for government agencies involved in the estuary. This approach would provide a forum for discussions and agreements on areas of fragmentation and overlap among agency responsibilities.
2. *A lead agency approach*, where key agencies would have taken the lead in decision making for overall and site- or species-specific issues. No agency would have lost

statutory authority, but all would have yielded specific responsibilities, and 'temporary' authority, to another agency for specific issue-by-issue application.
3. *An estuary council of governments*, a sort of 'super agency', that would have seen the establishment of a council of senior officials with executive authority consistent with existing statutes. Though agencies would have retained their authority, such a council would have been collaboratively staffed and funded and agencies would jointly direct its operations. This was a refinement of a recommendation from FRES phase one.

In the second phase, committee membership shifted to field-level personnel, who undoubtedly brought a different, more operational perspective. When it came to selecting an option, there was a preference for an approach that, while being strategic, also acknowledged jurisdictional tractability, could function within the historic context of inter-jurisdictional tension, and would best garner the commitment of implementing agencies. Each of the recommended approaches had integrative elements, but each entailed a shift in power arrangements. The provincial and federal governments decided to adopt a coordinated approach using what they then termed a *linked system*:

> In a linked system, management would continue to be entrusted to existing agencies which, to a large extent, would retain their present authority and responsibilities. However, instead of each operating separately, perhaps even in conflict with others, certain key agencies would be asked to cooperate in a joint process designed to improve on the present system but not replace it. The principle involved is to ensure that agencies work together to achieve a common set of goals—the estuary management policies recommended in the Management Program Report. (FRES 1982, 40)

With the creation of a new program, a setting would be provided for developing an integrated framework within which coordinated water/upland policy and planning could be developed and implemented. The FREMP was to emerge as a fusion of elements from the three possible approaches, where the final ingredients were chosen to maximize the potential for effective implementation.

In 1985 the implementation strategy developed by the multi-agency review committee formed the basis for a memorandum of understanding (MOU), an agreement between the provincial and federal governments to establish the FREMP. The first partner agencies were Environment Canada, Fisheries and Oceans Canada, the British Columbia Ministry of Environment (later called Water, Land and Air Protection), the Fraser River Port Authority, and the North Fraser Port Authority. The Greater Vancouver Regional District, akin to a county government, joined the program in 1991. The 1985 MOU had a five-year basis; it was renewed in 1991 and again in 1996, when the participating agencies made the FREMP an ongoing entity with no set time limit. The 1996 MOU also combined the FREMP and Burrard Inlet Environmental Action Plan and programs, administratively, but maintained a separate FREMP MOU.

This administrative merger was pushed by the regional district, which believed that one secretariat for the two programs would be 'more cost efficient'. The MOU was updated in 2003, but the basic workings of the FREMP remained intact.

Throughout the FRES process, there had been concern about what could be achieved within the political and administrative context. This concern extended to the delineation of the FREMP's boundary (Figure 7.1). The physical boundary encompasses about 155 km² of land on the wet side of the river's dykes, between about 50 km upstream and the outer banks of Boundary Bay and Point Grey—the marine estuary portion of the estuary. Despite an implicit understanding of many cause and effect relationships between land use and estuary water quality and upland activities and the stability of riparian habitat, the boundary is very limited, and this was deliberate. From the outset, it was believed that a broader boundary would heighten the potential for conflict among agencies, require realignments of authorities or jurisdictions, and perhaps even necessitate a change in existing power relationships. Instead, a 'tractable' boundary was established, one that conformed to the jurisdictions of the initial participating agencies, and it was decided that upland links could be developed through a management plan. After the Greater Vancouver Regional District joined the FREMP, the potential for creating these linkages was certainly enhanced. Overall, the FRES process had three main products: documentation of existing environmental conditions, a recommendation for a new integrated management and planning strategy, and an outline for implementing a preferred strategy. In essence, it was a softer version of the estuary council approach.

Operationally, formal powers remain invested in member agencies, though power is implicitly and sometimes explicitly exercised in a cooperative manner through the program. The framework for the FREMP is found in the MOU. The program's bureaucracy, now small (about four people), provides a support function for program committees (Figure 7.4) and serves as the program secretariat. The significance and influence of the FREMP model lie in its role as a venue for deliberative process and as a place for agency participation through public and private deliberation. It is a setting for the management of relationships. At one time the FREMP was a clearing house for information and research, and until the mid-1990s the program even facilitated and funded research, but this role has changed. At best, it now only manages to monitor estuary research and collaborate on events where the results of research, conducted by others, are presented.

As a planning entity, the FREMP focuses on enhancing the social significance and strength of technical/planning activities and reducing the potentially adversarial nature of inter-agency linkages. Implementation has evolved in terms of hierarchical integration, planning methods, responses to obstacles, and especially the allocation of resources. It is also aided by the adoption of five principles that seemingly acknowledge the complex nature of the relationships among the agencies involved and the agencies' diverse interests. These principles serve as a form of mission statement:[3]

1. Recognition of the importance of the estuary as a major economic and environmental resource in all planning and management activities;

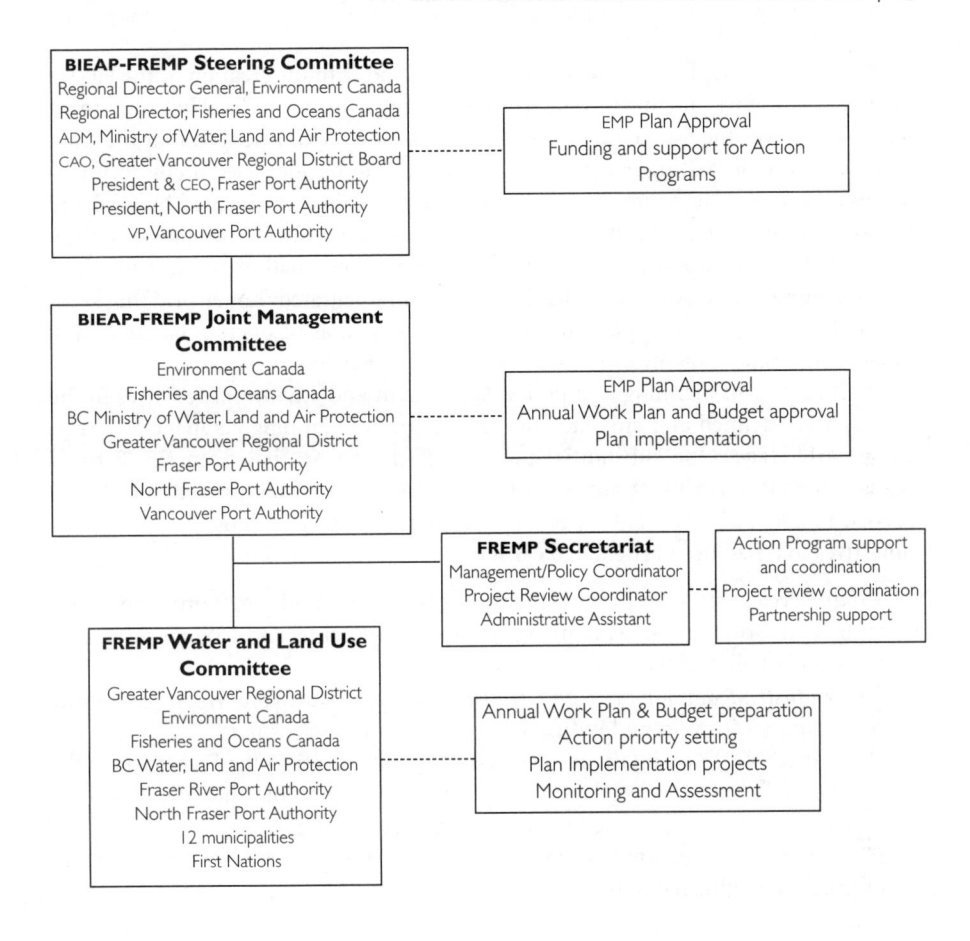

```
┌─────────────────────────────────────────┐
│ BIEAP-FREMP Steering Committee          │
│ Regional Director General, Environment Canada │         ┌─────────────────────────────┐
│ Regional Director, Fisheries and Oceans Canada│         │    EMP Plan Approval        │
│ ADM, Ministry of Water, Land and Air Protection│─ ─ ─ ─ │ Funding and support for Action│
│ CAO, Greater Vancouver Regional District Board│         │        Programs             │
│ President & CEO, Fraser Port Authority  │         └─────────────────────────────┘
│ President, North Fraser Port Authority  │
│ VP, Vancouver Port Authority            │
└─────────────────────────────────────────┘

┌─────────────────────────────────────────┐
│ BIEAP-FREMP Joint Management Committee  │
│ Environment Canada                      │         ┌─────────────────────────────┐
│ Fisheries and Oceans Canada             │         │    EMP Plan Approval        │
│ BC Ministry of Water, Land and Air Protection│─ ─ │ Annual Work Plan and Budget approval│
│ Greater Vancouver Regional District     │         │     Plan implementation     │
│ Fraser Port Authority                   │         └─────────────────────────────┘
│ North Fraser Port Authority             │
│ Vancouver Port Authority                │
└─────────────────────────────────────────┘
                    ┌─────────────────────────────┐   ┌─────────────────────────────┐
                    │     FREMP Secretariat       │   │  Action Program support     │
                    │ Management/Policy Coordinator│   │     and coordination        │
                    │  Project Review Coordinator │─ ─│  Project review coordination│
                    │   Administrative Assistant  │   │    Partnership support      │
                    └─────────────────────────────┘   └─────────────────────────────┘

┌─────────────────────────────────────────┐
│ FREMP Water and Land Use Committee      │
│ Greater Vancouver Regional District     │         ┌─────────────────────────────┐
│ Environment Canada                      │         │ Annual Work Plan & Budget preparation│
│ Fisheries and Oceans Canada             │─ ─ ─ ─ │    Action priority setting  │
│ BC Water, Land and Air Protection       │         │ Plan Implementation projects│
│ Fraser River Port Authority             │         │  Monitoring and Assessment  │
│ North Fraser Port Authority             │         └─────────────────────────────┘
│ 12 municipalities                       │
│ First Nations                           │
└─────────────────────────────────────────┘
```

BIEAP: Burrard Inlet Environmental Action Plan

EMP: Environmental Management Plan

FIGURE 7.4 Fraser River Estuary Management Program implementation structure

2. An emphasis on simplicity and practicality during implementation of the Program through the cost effective and shared deployment of existing resources;
3. Implementation of a management program in phases that allow for adjustment by the participating agencies;
4. Development of trust and consensus by ensuring broad consultation between participating agencies, by providing mechanisms for public consultation, and by maintaining program flexibility; and
5. Encouragement of compatibility between upland and foreshore uses. (FRES 1982)

Each of these principles has integrative elements, and with the creation of the FREMP the two senior governments recognized that no single agency or level has overriding jurisdiction over the estuary (Dorcey 1993). They also acknowledged that the fragmented nature of existing management would have to be addressed if environmental conditions were to be improved. Agencies were not just fragmented by level and jurisdiction; they also had divergent constituencies and any coordinated program that sought to manage the estuary on an integrated basis had to account for these relationships. The FREMP's approach is based on an integrated framework[4] that seeks the routine integration of planning and policy, as well as of operational activities, notably environmental reviews, among its partner agencies.

With respect to planning and policy, key human and environmental uses in the estuary were defined and embedded in the program's planning language at a relatively early stage. One substantial product of the FREMP has been the Estuary Management Plan (EMP) (cf. FREMP 1994 and 2004). The EMP was created in 1994 and revised in 2003. The first EMP specifically sought to balance a range of use-related objectives that had been articulated earlier:

- *Urban development*—to provide the means for accommodating a growing population and economy while maintaining the environmental and ecological productivity and quality of the estuary.
- *Port industrial development and transportation*—to provide viable economic opportunities for port and industrial development in the estuary.
- *Water quality*—to maintain ambient water-quality levels in the Fraser River, the outer estuary, and adjacent bays to ensure the preservation of fisheries and wildlife and, where possible, provide for water-contact recreation.
- *Habitat management*—to maintain and, where feasible, increase the productivity of fish and wildlife habitat.
- *Recreation*—to enhance the quality and variety of estuarine areas that have recreational qualities. (FRES 1982; Kennett and McPhee 1988)

The EMP now includes seven action programs—integration, water and sediment quality, fish and wildlife habitat, navigation and dredging, log management,[5] industrial and urban development, and recreation. With the exception of the integration plan, these programs mirror the 'area designations' that appear in other FREMP planning documents. Area designations were negotiated between FREMP partners and the municipalities in order to link the FREMP's shoreline habitat classification (based on coding habitat areas according to environmental productivity) with upland zoning and planning. The classification system is used in the environmental review process (described below). In the current EMP, the integration action plan seeks to better link the EMP with upland (local government) planning processes, largely through what is termed a 'features and functions approach to management and decision making.' This approach is based on identifying important social, economic, and environmental attributes of the river's reaches and then identifying areas where consensus between the various plans exists and where it does not (FREMP 2004, 31). The development of explicit foreshore (FREMP) planning in conjunction with local government planning has long been an objective of the FREMP. It is telling that building such links remains an objective in the 2003 EMP. Despite the best intentions, integration of FREMP objectives with municipal plans has been difficult, even hesitant, and remains so.

Integration of environmental reviews is facilitated through the Coordinated Project Review Process (CPRP). This process coordinates what was previously a cumbersome arrangement of multiple-agency review and approval procedures for estuary development proposals. With the CPRP there is one review body—the Environmental Review Committee (ERC)—where relevant agencies review applications. The system of habitat classification that developed through the FREMP helps guide the review process. But the ERC does not approve or reject projects. That responsibility remains with the 'lead agencies', which for foreshore development applications most often turn out to be the port authorities. The ERC provides a recommendation for the lead agency responsible for making a formal decision about the project. The environmental review occurs within the integrated setting, but the authority for development approval, or rejection, remains vested in partner agencies.

Public consultation is sporadic and issue specific. At best, members of the public and organized environmental groups are peripheral to the decision-making processes. Non-government involvement is exercised largely through information events. This limited participation model may be at odds with the ideal interpretation of an integrated approach, but as with the limited mandate and boundary, limiting non-government participation was part of the implementation strategy.

DISCUSSION AND CONCLUSIONS

The Fraser River estuary illustrates some of the challenges encountered in putting an integrated approach to resource and environmental management into practice. Implementation occurs at different management levels and across jurisdictions. Negotiation, compromise, and shifting goals have affected the FREMP process. Considering these factors, it is not unexpected that progress in implementing some of the program's elements has been slow. What has emerged is a hybrid program that represents elements of administrative and bottom-up approaches, where both hierarchical integration and the input of ground-level practitioners have been influential in the program's design and implementation.

Tractability

The FREMP could not have been implemented, and would certainly not be as long-lived, without the support, resources, and authority of the existing agencies. Such support has been forthcoming because the program was designed with an explicit acknowledgment of implementation issues, in particular the potential for conflict should a program requiring a substantial shift in power arrangements be advanced. The translation of program language into practical plans and the application of logistical support are facilitated by a supportive bureaucracy. By working within the context of existing power arrangements, the program minimizes the potential for tractability problems. There was also an early focus on defining management and physical boundaries in terms that would maximize support from the founding agencies. But the boundaries could also be expanded as relationships with local governments were developed and linkages to upland planning established. The emphasis has been on coordinating agencies to integrate planning efforts, assembling information on environmental conditions to guide planning, and integrating the review of shoreline development applications. Each of these elements was seen as a practicable objective when the program was being developed, and each has emerged as having what can be termed 'a clear value to the agencies working in the river, that's why they support FREMP.'[6]

While the FREMP facilitates communication among interests and provides for the exchange of views and values among decision makers, it is far from being an inclusive model. Not all affected groups are an explicit part of decision-making processes, although this does not mean that conservation organizations and other interests have not been influential. Agencies bring to the FREMP process the experiences and knowledge gained from interaction with their constituents through formal and informal participation settings. These indirect venues have demonstrated a capacity for influencing the way the FREMP operates (Hanna 2000).

Clarity

Tractability and clarity (with respect to program objectives) have evolved in tandem. At the end of the FRES process, it was recommended that an estuarine management program be developed. And by the second phase of the FRES, it had been decided that the resulting program would have basic facilitative and information-parsing functions. The shift in involvement, from senior public servants in the first FRES phase to field-level managers in the second, supported a bottom-up approach to program design and likely helped ensure that a stronger operational perspective was incorporated into the FREMP. There was clarity about what the FREMP would not be. It would not be a new super-agency. Power and authority structures were to be maintained, and the new program would serve as a clearing house for actions and as a place where conflict could be addressed. It would also evolve as a setting for the building and establishment of relationships. The enabling framework was structured in a way that helped maximize the potential for long-term agency support and the eventual inclusion of other agencies and local government in the planning process. The objectives were cooperative and tractable.

Knowledge, Commitment, and Information

Communication, interaction, and information development have been systematically exercised within established political and bureaucratic structures. But the FREMP process has expanded the base knowledge of environmental quality and trends, and this in turn has influenced the planning process. Planning and public policy implementation are often conflictual processes directed towards building consensus. Implementation has been accomplished in part through the shaping of attention, where information is shared among agencies. This in turn affects organization by creating a collective inter-agency identity where shared perceptions of issues and solutions become action realized through the program.

While information development was an important initial function, the program's emphasis is on planning and environmental review rather than on enforcement and regulation. Before FREMP's inception, planning and environmental management in the Fraser River estuary was fragmented, ad hoc, and reactive. The creation and application of provincial and federal policies often occurred in jurisdictional isolation. The FREMP's primary accomplishment is its integrative role. This is demonstrated through the coordinated planning (the EMP) and review system (CPRP), which has in turn contributed, albeit indirectly, to better management of the estuary's environment (Hanna 2000). The planning products are responses to the pre-program conditions. The extent to which these have affected the behaviour of individual agencies is variable, but the program has improved awareness of the multiple-use objectives of other agencies and, by proxy, of their constituencies. The partner agencies' commitment is shown through the longevity of the FREMP and the agencies' willingness to support it even during years of fiscal restraint. The initial MOU has been renewed and expanded, and the FREMP is now an ongoing program.

Personalities

At the political level, the relationship between the federal government and British Columbia is sometimes an irritable one, and especially so with respect to resource and environmental management. While there have been issues of disagreement in the estuary, federal/provincial interaction at the bureaucracy level has been amiable. This has been an important factor in both the initiation and implementation of the FREMP. A recurrent theme in discussions with agency personnel, regardless of their level within an agency, is the extent to which the FREMP process has benefited, and reflects, a collegial environment. As one senior public servant commented in earlier interviews, the people who 'sit at the FREMP table like each other and this has made it possible to tackle some tough issues.' While this collegiality has supported integration efforts, it can also be an unstable dynamic, dependent on the continuity of personnel and their ability to maintain a certain level of support from their own organizations. It is difficult to measure this aspect definitively, although we can see that membership on FREMP committees has changed over the last decade and that some of those from the formative years of the program who possessed a strong knowledge of institutional history and commitment have moved on. As committee memberships have changed, so have the resources allocated to the FREMP.

History and Unique Challenges

In the estuary there was, and some argue there remains, a legacy of conflict. But there is also a history of cooperation, of sharing planning resources, and of strong interpersonal and inter-agency relationships. In the FREMP case, history has been an important aspect of implementation and was certainly a consideration in the program's design.

A unique aspect of implementation in the Fraser estuary has been the relative support of the political versus the bureaucratic levels. The FREMP is largely a product of the public service. It is not uncommonly assumed that the vision for integrated approaches begins in the political arena and that bureaucrats are by nature opposed to inter-agency cooperation. While this can be the case in some settings, in others there are long legacies of inter-agency synergy. In the FREMP situation, the political level has been only peripherally involved. The integrative approach embodied in the FREMP has evolved despite a lack of political interest, or perhaps because of it.

The approach to integration represented by the FREMP has worked well in addressing the challenges of implementation. There is a concern that the program, in addressing the implementation imperative, has taken a path of least resistance rather than develop the strong planning and management initiatives that some see as being needed to ensure the integrity of the estuary's ecosystems. In part this is true. Given the history of intergovernmental conflict, the complexity of uses, and the number of agencies with an interest in the estuary, a more intrusive and authority-imbued program approach would have been resisted and would ultimately have been short-lived. The FREMP does not threaten agencies. Tractability and the role of the program statute in framing the implementation process have been defining aspects of the program's design, and the program ensures that existing formal power relationships remain intact, which is in part why the program enjoys continued institutional support. But this approach has yielded cautious planning, and it seems that some early objectives, notably the development of foreshore and upland planning links, remain ongoing—some might say unfulfilled. The problem-solving approach is based on what Forester (1989) characterizes as professionalism embedded within an institutional context, but this was intentional. The application of technological and regulatory approaches is indirect and is exercised through government stakeholders. Political argument, participation, and policy alternatives that might conflict with existing patterns of power are restricted by the nature of the MOU that established the FREMP. The location of power has become more diffuse than is initially apparent, and the FREMP may not be quite the toothless tiger that some might see it as.

The model of integration created by the FREMP has resulted in a new set of rules for interaction, one where power is no longer habitually exercised by agencies individually, but within a collaborative setting. Agencies may not always be explicitly aware of these new rules of deliberation and action, but their support for the program, their deference to it as an avenue for problem-solving, and their support for program enhancement all suggest an undeclared and unrecognized exercise of power through the program. With careful attention to implementation, even a modest approach to integration can lead to shifts in the way agencies exercise power towards cooperative and more effective natural resource and environmental management.

NOTES

1. Support for this work was provided in part by the Canada Mortgage and Housing Corporation and the Soil and Water Conservation Society through the K.E. Grant Fellowship.
2. The case discussion is based on a study completed in 1998 that included discussions with FREMP staff and partner agencies. Recent (2005) discussions with program staff and a review

of current planning documents are also incorporated into this chapter. Earlier results related to environmental quality and planning are discussed in Hanna 1999 and 2000.
3. These were outlined in the second phase of the FRES.
4. The framework for managing the estuary has also been characterized as a sustainability approach (e.g., Dorcey 1993).
5. This warrants a brief explanation. Logs towed from sites on the Pacific coast or from upriver are stored in the estuary before being processed. Log storage accounts for a sizable area of estuary surface use. Given the proximity of log storage areas to shoreline habitat, this activity has had significant impacts on riparian systems. Changes to log storage practices have been a key part of improving shoreline habitat conditions.
6. A comment made by a senior public servant from a FREMP partner agency.

REFERENCES

Baker, R. 1989. Institutional innovation, development and environmental management: An 'administrative trap' revisited, Part I. *Public Administration and Development* 9:29–47.

Brekke, J.S. 1987. The model-guided method for monitoring program implementation. *Evaluation Review* 11 (3): 281–99.

Coleman, W.D. 1988. *Business and politics, a study of collective action.* Montreal and Kingston: McGill-Queen's University Press.

Dorcey, A.H.J. 1993. Sustainable development of the Fraser River Estuary, Canada: Success amid failure. *Coastal zone management, selected case studies.* Paris: OECD.

Dryzek, J.S. 1990. Designs for environmental discourse: The greening of the administrative state. In R. Paehlke and D. Torgerson (eds), *Managing leviathan.* Peterborough, ON: Broadview.

Forester, J. 1989. *Planning in the face of power.* Berkeley, CA: University of California Press.

Freeman, H. 1980. The present status of evaluation research. In *Evaluating social action projects,* 9–46. New York: UNESCO.

FREMP (Fraser River Estuary Management Program). 1994. *A living working river: An estuary management plan for the Fraser River.* New Westminster, BC: Fraser River Estuary Management Program.

——— 2004. *A living working river: An estuary management plan for the Fraser River.* Burnaby, BC: Fraser River Estuary Management Program.

FRES (Fraser River Estuary Study). 1982. *A living river by the door, a proposed management program for the Fraser River estuary,* FRES Phase II. Victoria and Vancouver: Province of British Columbia and Government of Canada.

Ham, C., and M. Hill. 1993. *The policy process in the modern capitalist state.* Toronto: Harvester Wheatsheaf.

Hanna, K.S. 1999. Integrated resource management in the Fraser River estuary: Stakeholder's perceptions of the state of the river and program influence. *Journal of Soil and Water Conservation* 54 (2): 490–8.

——— 2000. The paradox of participation and the hidden role of information: A case study. *Journal of the American Planning Association* 66 (4): 398–410.

Hessing, M., and M. Howlett. 1997. *Canadian natural resource and environmental policy.* Vancouver: University of British Columbia Press.

Hyder, M. 1984. Implementation: The evolutionary model. In D.L. Lewis and H. Wallace (eds), *Policies into practice: National and international cases studies in implementation*, 1–18. London: Heinemann.

Jones, C.O. 1984. *An introduction to the study of public policy.* Monterey, CA: Brooks Cole.

Kennett, K., and M.W. McPhee. 1988. *The Fraser River estuary: An overview of changing conditions.* Vancouver: Fraser River Estuary Management Program.

Linder, S.H., and B.G. Peters. 1989. Instruments of government: Perceptions and contexts. *Journal of Public Policy* 9:35–58.

McPhee M.W., and J.D. Wiebe. 1986. Coordinating management activities in the Fraser River estuary. In R Lang (ed.), *Integrated approaches to resource management*, 229–50. Calgary: University of Calgary Press.

Margerum, R.D. 1999. Integrated environmental management: The foundations for successful practice. *Environmental Management* 24 (2): 151–66.

Margerum, R.D., and S.W. Born. 1995. Integrated environmental management: Moving from theory to practice. *Journal of Environmental Planning and Management* 38 (3): 371–90.

——— 2000. A coordination diagnostic for improving integrated environmental management. *Journal of Environmental Planning and Management* 43 (1): 5–21.

Mitchell, B. 1986. The evolution of integrated resource management. In R. Lang (ed.), *Integrated approaches to resource management*, 13–26. Calgary: University of Calgary Press.

——— 1990. Integrated water management. In B. Mitchell (ed.), *Integrated water management: International perspectives and experiences.* New York: Bellhaven.

——— 2002. *Resource and environmental management.* Harlow: Prentice Hall.

Paehlke, R. 1990. Democracy and environmentalism: Opening a door to the administrative state. In R. Paehlke and D. Torgerson (eds), *Managing Leviathan*, 35–58. Peterborough, ON: Broadview.

Platt, J. 1973. Social traps. *American Psychologist*, August.

Rayner, S. 1991. A cultural perspective on the structure and implementation of global environmental agreements. *Evaluation Review* 15 (1): 75–102.

Rossi, P., and H. Freeman. 1985. *Evaluation: A systemic approach.* Newbury: Sage.

Sabatier, P.A., and D.A. Mazmanian. 1981. The implementation of public policy: A framework for analysis. In D.A. Mazmanian and P.A. Sabatier (eds), *Effective policy implementation.* Lexington: D.C. Heath.

Skogstad, G., and P. Kopas. 1992. Environmental policy in a federal system: Ottawa and the provinces. In R. Boardman (ed.), *Canadian environmental policy: Ecosystems, politics, and process*, 43–59. Don Mills, ON: Oxford University Press.

Torgerson, D. 1990a. Obsolescent leviathan: Problems of order in administrative thought. In R. Paehlke and D. Torgerson (eds), *Managing Leviathan*, 17–34. Peterborough: Broadview.

——— 1990b. Limits of the administrative mind: The problem of defining environmental problems. In R. Paehlke and D. Torgerson (eds), *Managing Leviathan*, 115–64. Peterborough: Broadview.

Van Meter, D.S., and C.E. Van Horn. 1975. The policy implementation process, a conceptual framework. *Administration and Society* 6(4): 445–88.

Walther, P. 1987. Against idealistic beliefs in the problem solving capacities of integrated resource management. *Environmental Management* 11(4): 439–46.

Weale, A. 1992. Implementation failure: A suitable case for review? In E. Lykke (ed.), *Achieving environmental goals: The concept and practice of environmental performance review*, 43–63. London: Belhaven.

CHAPTER 8

Resolving Human–Grizzly Bear Conflict: An Integrated Approach in the Common Interest

Seth M. Wilson and Susan G. Clark

INTRODUCTION

Resolving human-wildlife conflict in the common interest is problematic, especially when grizzly bears (*Ursus arctos*) conflict with humans at the interface of public and private lands. The unwelcome activities of bears cover a broad spectrum, from killing livestock or destroying beehives to foraging for garbage close to homes. In rare cases, bears threaten human safety. Low-elevation private lands adjacent to public lands are often areas of conflict. Available attractants from human activities found on these lands can coincide with critical spring and fall grizzly bear habitat. Repeated incidents or conflicts often lead to bear habituation, food conditioning, and removal. Human-caused mortality jeopardizes conservation gains and makes long-term bear viability difficult. One way to improve bear survival and enhance their distribution is to prevent conflicts. Grizzlies have been given protection under the US Endangered Species Act because of historic and ongoing human impacts that are expected to continue unless a concerted effort is made to minimize them. Our work targets this problem.

Conflicts are certainly a technical problem, but they are also a human one. Many are caused by people whose practices inadvertently draw bears into problem situations. Others are issue-based, resulting from people's different views on the status of bears, causes of the problem, and what should be done (Primm 1996; Servheen 1989). Consequently, 'the single most important variable . . . is likely social not biological' (Mattson et al. 1996, 155). We believe that managing conflict successfully requires an integrative approach that goes far beyond biology alone. We know that maintaining bears on private lands requires active, constructive interaction among ranchers, non-agricultural residents, wildlife managers, industry representatives, and conservationists in innovative partnerships (Primm and Wilson 2004; Clark and Brunner 2002). Partnerships organized around research and conservation are one means to address the challenge. Integrated methods that ensure the rationality, practicality, and morality of the problem-solving efforts enhance opportunities for success, regardless of the problem at hand (Clark et al. 2001). Our approach to the multi-faceted human-bear conflict situation uses this approach.

We participated in two case studies whose purpose was to improve understanding and reduce conflict: Rocky Mountain Front and Blackfoot Valley, Montana (Figure

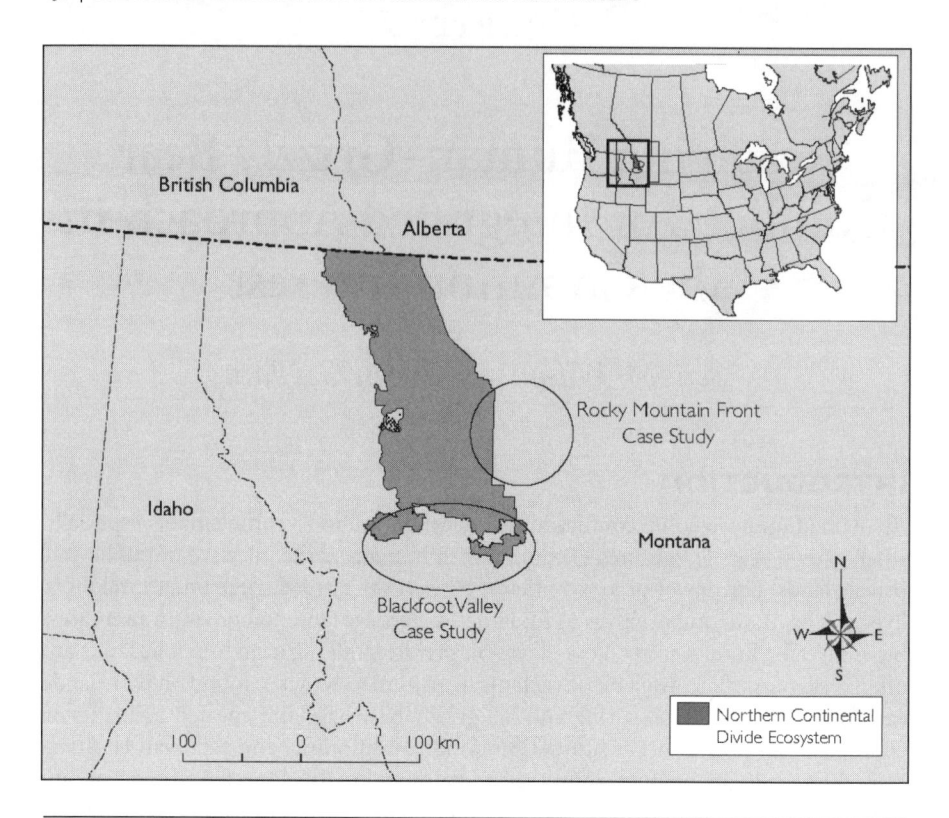

FIGURE 8.1 Location of Rocky Mountain Front and Blackfoot Valley case studies in Montana, USA

SOURCE: Data layers from Natural Resources Information Service, http://nris.state.mt.us.

8.1). This chapter (a) describes the context of our work and methods and defines the problem, (b) analyses the two cases, and (3) offers lessons for integrated problem-solvers, well beyond grizzly bear conservation. The integrated approach we used is problem oriented, contextual, and multi-method (Clark et al. 2001). The traditional approach to large carnivore management that currently dominates is fragmented, bureaucratic, and deeply institutionalized, but it is slowly and reluctantly moving towards a more integrated approach.

CONTEXT, METHODS, AND PROBLEM DEFINITION

Conflict between humans and grizzly bears occurs within unique contexts. Understanding context is central to resolving conflict in the common interest—that is, in the interests of the majority of people (Brunner et al. 2002). In contrast, serving only special interests benefits the few at the expense of the majority. We studied the context of human-bear interaction and produced a social and biological definition of the conflict problem (Clark et al. 1996).

Integrated Approach

Our approach is integrative in several senses. First, we used a logical and comprehensive framework that invites integration of the biological and social sciences for problem-solving (Clark 2002a). Second, we integrated empirically derived case material about human-bear conflicts into the framework in order to examine the relationships among the variables studied and to check what is known and unknown about such conflicts and their context against recommended standards for decision making (Lasswell 1971). Third, we sought integrated solutions that require the discovery of a new mode of cooperation (Clark and Mattson 2005). Such integrated solutions go well beyond winners and losers or mediocre compromises. Integrated solutions require redefining the context so that conflicting parties can satisfy their underlying-value demands. This approach leads to new practices, such as 'prototyping'(Clark et al. 2002a). We understand management and policy improvement to mean upgrading the outcomes of the decision process to approximate established standards.

Context

Some bear managers, landowners, and conservationists already work contextually (e.g., Madel 1996; Jonkel 2002; Primm and Murray 2005; McLaughlin et al. 2005). The management and planning efforts of the Montana Department of Fish, Wildlife and Parks (FWP) have been contextual and effective, and exemplify a strong commitment to bear conservation that incorporates values shared by local rural communities while representing national interests under the Endangered Species Act (Dood et al. 2006). The context is the set of conditions that shapes problems and solutions and can be thought of as a human 'social process' or 'envelope' (Taylor and Clark 2005). Muth and Bolland (1983) note that the social process surrounding any problem involves people seeking to maximize gratifying outcomes by using society's institutions to influence the distribution of resources in their favour. In our work, we studied the context of human-bear conflict using anthropological and other social science methods that allowed us to identify who was participating, reveal their perspectives, detail the situation or arena of interaction, ecologically and socially, see the values at stake or in play for the people involved, ascertain their strategies, and understand the outcomes they sought.

In grizzly bear conservation, it is the human social process that conflicts with bears. As a result, the solution lies in changing the context. The present spatial and temporal aspects of human-bear interactions are dependent on historic and existing social processes—human perspectives and practices. These contextual factors are key elements in managing bear population viability, conflict, and mortality risks. Lack of contextual problem-solving is common in management, especially when scientific management and bureaucracy dominate (Clark et al. 1989; Brunner 1991).

The issues at stake for participants in our cases were similar, but they varied with the individual and situation (Wilson 2003; Primm and Wilson 2004). Everyone wants to live with dignity, and this means having adequate values. Included in what people value are power, enlightenment, skill, wealth, well-being, affection, respect, and rectitude (Clark and Wallace 2002). All people want to be able to shape (produce, give) and share (receive, use) all eight to the fullest extent possible. Collectively, these values represent security. Conflict with bears threatens people's security, but in different ways. The way that people and their values interact lead to conflict in the first place and later to the ability to reduce conflict, as described below. In both of the cases we studied, because of our participation and the work of other people, the value positions of most participants improved and conflicts were reduced. In this sense, outcomes were integrative.

Methods

We used case study methods guided by the analytic framework introduced above to organize our work (Clark and Willard 2000). In both cases, Wilson carried out all fieldwork, data collection, and technical analysis. This approach requires users to be explicitly and systematically problem oriented, contextual, and multi-method, as described below. 'Like any good cartographic tool, the framework helps us map a route through a "policyscape" to a destination, highlighting key features and revealing both pitfalls and opportunities' (Clark 2002a, 9).

Problem orientation is a strategy to address problems. It is about being procedurally rational. Five tasks help people orient themselves effectively to the problem at hand. These tasks can be easily appreciated if we ask five questions:

- What are our goals?
- What historic trends should be described to evaluate whether goals are being met?
- What are the conditions behind these trends?
- What will happen in the future if trends and conditions remain unchanged?
- If we find we have a problem or discrepancy between goals and the future, what can be done about it?

In our two cases, we helped the people involved orient themselves more effectively towards human-bear conflicts. Our common interest goals were to conserve bears in ways that minimize conflict and engender broad-based, long-lasting public support. With this goal, we sought to identify the trends and conditions that led to repeated conflicts on private agricultural lands. We worked closely with landowners and others and found opportunities to improve understanding among researchers, managers, and conservationists about conflicts between livestock producers and bears. In essence, we gathered, processed, and disseminated information about trends, conditions, and projections. In the Rocky Mountain Front case, we left implementation to citizens and wildlife managers, and in the Blackfoot Valley case, we helped implement options that we are now monitoring.

Being contextual means being cognizant of the social and decision-making processes at play. In fact, by definition a management or policy improvement is about upgrading the decision process. Decision making is concerned with what people do, as well as with what they say and expect. In human-bear conflict, the kind of decision process in place determines how the conflict problem is managed. The manner in which the decision-making process unfolds determines whether conflict or other problems can be understood usefully and whether win-win solutions are possible as outcomes (see Weiss 1989). The conflict problem and the associated decision process must be related to the social process that envelops decision making if a solution is to be found. Decision making is made up of interrelated functions or activities. These are usually ongoing, all at the same time. All of them must work well in order to for the common interest to be served. In brief, decision functions may be described as follows (Clark and Brunner 2002):

- Intelligence is about information essential to decision-making quality.
- Promotion is about debate, advocacy, and urging proposals or possible solutions.
- Prescribing is about setting or choosing plans.
- Implementation is about putting prescribed plans into action on the ground.
- Appraisal is about monitoring and evaluating performance.
- And termination is about ending the effort or arrangement and moving on to another effort or adjusting the old effort.

Common interest standards have been recommended for each function and for the overall decision process (Clark 2002a, 60, 71–6). These are the terms and standards that we have used.

In our cases, we sought to upgrade all functions as opportunities occurred. We wanted to contribute to the pre-existing efforts of the Montana Department of Fish, Wildlife and Parks and look for opportunities to help, collaborate, and enhance efforts when possible. For example, we sought to make the intelligence about human-bear conflict as factual, open, and comprehensive as possible. We sought to make promotion as rational, integrative, and effective as possible. We sought to make the prescription as balanced, inclusive, and future directed as possible. We sought to make implementation timely, non-threatening, and in the common interest. We sought to make appraisal as unbiased, contextual, and constructive as possible. And finally, we sought to make the termination as timely, comprehensive, and ameliorative as possible. We were comprehensive (i.e., we included all relevant people), as noted earlier, and also selective in targeting participants (e.g., ranchers and residents) whose practices most affect bears. We had to understand the spatial and temporal situation of conflict. We focused on understanding how each decision function played out in actual circumstances and how landowners figured into them, function by function and standard by standard. We were especially interested in learning about the human behaviour and management practices that are a product of the context in which humans and wildlife live and coexist. Thus, we sought to account for human social and ecological factors that predispose bears to come into conflict with humans (Clark et al. 2002a).

In our cases, we focused on three biophysical trends and conditions that directly reflect the social and decision-making processes involved: (a) historic and current livestock management and beekeeping practices; (b) ecological features that provide habitat for grizzly bears and bring them close to humans; and (c) the spatial and seasonal distribution of reported and verified human-bear conflicts. We used geographical, statistical, and database software capable of displaying, organizing, and analyzing data sets to map and display these trends and conditions (Wilson 2003).

Problem Definition

In wildlife conservation, the problem of conflict is often construed to be only a biological problem (Clark 1997). In contrast, our definition of the problem shows that conflict originates in the perspectives and practices of people. Bears only respond to easily accessible food that people inadvertently make available. Grizzly bear managers from FWP have known this for decades and have been working to change people's perspectives and practices. We shared this view of the problem and felt that we could contribute to the department's ongoing efforts by finding ways to change the location, abundance, and availability of attractants, thereby reducing opportunities for conflict. In short, our problem definition was a contextual and people-centred one. We defined the conflict problem to capture the human social dynamic for the people involved. This required developing a problem definition that could be easily understood and that made both social and biological sense. According to Weiss (1989), by defining problems, we impose a framework on troublesome situations. Definitions suggest the causes and consequences of the problem and propose a theory for solution. The definition in fact configures its solution.

BACKGROUND OF ROCKY MOUNTAIN FRONT AND BLACKFOOT VALLEY CASES

In both case study locations, the Montana Department of Fish, Wildlife and Parks has active and effective management programs that focus on preventing conflicts by involving local people. Indeed, the willingness of FWP to collaborate with us represents the progressive and professional nature of this agency and its grizzly bear managers. The account of our work in this chapter is merely an attempt to formalize how we approach problem-solving and how we tried to contribute to the existing and positive efforts of FWP. Our involvement on the Rocky Mountain Front began in 1997 and ended in 2003, and our work in the Blackfoot Valley began in 2002 and is ongoing. Before describing these cases, we should more fully explain the problematic situation of human-bear conflict.

The Problematic Situation of Human-Bear Conflict

Both humans and bears face a problematic situation in areas where they come into contact. Historic settlement patterns in creek and river bottoms in the northern Rocky Mountains and the subsequent private ownership of these lands have had important consequences for grizzly bears, who depend on these same riparian habitats for seasonal and life history needs.

In terms of people and their agricultural activities, we found that the region is dominated by cow/calf operations characterized by winter feeding, centralized and spatially fixed operations, irrigated hay production, and docile breeds of cattle (Dale 1960;

Jordon 1993). Livestock management practices overlap with the grizzly bears' emergence from their dens in the early spring. Bears routinely visit calving areas, and the traditional practice of dumping dead livestock into spatially fixed boneyards (carcass dumps) is partly to blame for chronic livestock–grizzly bear conflicts (Wilson 2003).

We also found that the State of Montana's apiary licensing system, which originated in the 1940s and supported commercial beekeeper 'territories', increases the likelihood that foraging bears will find beehives by encouraging the broad spatial distribution and use of beehive sites. The state requires that beehive sites be registered to the quarter section and used annually. Irrigated alfalfa typically grown in riparian areas produces premium honey. Honeybees, known as 'nectar robbers', produce honey surpluses that beekeepers harvest. Alfalfa cropping results in pulses of surplus honey associated with the flowering cycle. Beehive sites near irrigated alfalfa, riparian areas, and prairie grasslands are thus ideal locations for producing honey, and they coincide with bear habitat.

The combination of the trends in cow/calf operations and beehives creates conditions for conflict. Given this, we were surprised to learn that since the listing of grizzly bears as a threatened species in 1975, no one had contextually and systematically inventoried attractants (calving areas or beehives) in an agricultural landscape and engaged with private landowners to learn more about their land-use practices. Nor had the spatial locations of verified conflicts or incidents between bears and landowners been mapped or used explicitly in management. Given the historical conditions that give rise to human-bear conflicts and the lack of systematic research on current conditions, we felt that it would be productive to work with state grizzly bear managers, private landowners, and local conservation groups to achieve better understanding of the problem and help find ways to avoid future conflicts with bears.

For bears, the picture has been a dramatic population decline historically. More recently the small remnant populations have expanded into agricultural areas. Population decline is linked with human causes, as human-bear conflicts are often a precursor to mortality. Over the past 20 years in southern Canada and the United States, 85 per cent of known bear mortality has been attributed to humans (Mattson et al. 1996). In states like Montana, grizzly bear mortality is spatially concentrated on the periphery of occupied core habitats. For example, in north-central Montana, where both case studies are located, the US Fish and Wildlife Service (2003) estimates that approximately 60 per cent of known and probable grizzly bear deaths occur on or within 1 km of private land. These less secure, low-elevation habitats are typically privately owned agricultural lands, contain a variety of unnatural bear foods, and yet are important to bear conservation. Since bear dispersal is a slow process, conservation efforts must recognize that minimizing the lethality of these habitats for bears in these areas is critical (McLellan and Hovey 2001).

Current management efforts are organized in ways that both aid and limit resolution of conflict. Organizational arrangements for bear recovery under the auspices of the Interagency Grizzly Bear Committee (IGBC) primarily focus on coordination of management activities among natural resource agencies on public lands. Thus, state wildlife managers have played a prominent role in conflict management on private lands. However, they have not always had the time or resources to involve private landowners, local conservation groups, or independent researchers directly in systematic, coordinated, and integrative efforts that can aid conflict management. Managers frequently have to respond to volatile situations where irate landowners and ranchers are upset over property damage and perceived or genuine threats to human safety. They often find themselves in a mode of reactive prevention, where specific techniques, such as bear-resistant garbage containers or electric fencing around beehives, are only employed following conflicts (Wilson 2003).

This pattern is amplified when conflicts involve suspected livestock depredation by grizzly bears. This triggers an investigation to verify possible depredations by the US Department of Agriculture's Animal, Plant, and Health Inspection Service (Wildlife Services), an agency whose fundamental task is to protect American agriculture. Both Wyoming and Montana have memorandums of understanding (MOUs) with Wildlife Services to co-investigate livestock-related conflicts. During co-investigations with state wildlife managers, Wildlife Services often handles the trapping of 'problem' bears, which are then relocated by the state. While Wildlife Services does not share cooperative decision-making authority under IGBC guidelines and has no official representation on any IGBC committees, its manner of addressing conflicts represents a management philosophy starkly divergent from one that would promote the protection and conservation of bears overall. Its actions overemphasize control of bears by trapping, relocating, or even killing. This approach contrasts with that of FWP, which focuses on changing human behaviour to avoid or minimize conflict in the first place. Some landowners and ranchers find themselves caught between the differing management philosophies of the two branches of government, each having different perspectives on the conflict problems and their solutions.

Additionally, the US Fish and Wildlife Service, FWP, and dozens of conservation groups support bear population expansion and improved distribution to restore genetic and demographic connectivity among presently isolated populations (Servheen 1998; Dood et al. 2006; Finkel 1999). Despite laudable efforts to enhance connectivity potential for bears through habitat protection and identification of highway linkage zones, the US Fish and Wildlife Service has not adequately integrated private landowners or ranchers into the recovery process (Primm 1996). Finally, while state bear managers in Montana have made invaluable contributions to grizzly recovery in an understaffed and under-budgeted agency, they often simply lack the time to explore other management options. Other options include undertaking integrative research, involving non-agency researchers, and capitalizing on the existing efforts of local, grassroots conservation groups to find common ground that benefits all interests. Without collaboration, these trends and conditions will likely continue. We chose to participate and contribute via two prototyping efforts.

Improving the Decision-Making Process as a Management Outcome

Improving management by reducing conflicts means upgrading the present decision-making process. The goal of bear management, and our goal, is to minimize conflict in ways that have the enduring support of the public and are in the common interest. This means balancing the interests of local and national publics and addressing legitimate local conflicts. In our cases, the individual participants varied, histories of conflict differed, and people were taking different but limited actions to prevent conflict. In both cases, we helped to organize a more effective human social process and to upgrade decision making in response to conflict. We did this through our work, relationships, and communication. We targeted the intelligence, debate, guidelines or norms, implementation, monitoring, and evaluation functions, and adapted the programs over time. We sought to approximate recommended standards—for example, by having integrity and being factual, reliable, comprehensive, open, fair, and honest. We evaluated both process and substantive outcomes in a never-ending, active, learning way (Clark 2002b). Our interventions had positive outcomes (described below) that improved land-use practices, communication and deliberation among ranchers, their management decisions, and their monitoring of their own activities. Finally, preliminary results suggest that there has been a reduction in the number of human–grizzly bear conflicts.

ROCKY MOUNTAIN FRONT CASE

The foothill grasslands of the Rocky Mountain Front represent the only place in the lower 48 United States where grizzly bears have continuously occupied their historic prairie habitat (Aune and Kasworm 1989). Yet the presence of grizzly bears has been controversial. Conflicts between bears and residents began in the mid-1980s, peaking by the mid- to late 1990s. They have slowed in the 2000s, though have periodic spikes, depending on seasonal availability of natural bear foods. Grizzly bears have been and still are both reviled and revered by local residents. In many respects, it is remarkable that bears have maintained a subpopulation along the Rocky Mountain Front. This situation represents an excellent opportunity for people in all relevant areas to study and improve the interactions of agriculturalists and a large carnivore. Since the area is dominated by livestock ranching and honey production and since the bulk of conflicts are related to attractants, we first wanted to establish the relative spatial associations among landscape features, attractants, and the likelihood of conflicts by focusing on the locations and management of cattle, sheep, and beehives.

Case-Specific Methods

How did we go about trying to improve the outcomes of the social and the decision-making processes and minimize conflict? We used in-person interviews, Geographic Information Systems (GIS), and standard statistical software for data collection and analysis. GIS has been a particularly useful tool in involving local people in the research and conservation process, and our efforts reflect our conviction that local knowledge can be appropriately used in rigorous scientific studies (Berkes et al. 1998; DeWalt 1994). Aspects of all these methods were also used in the second case study, the Blackfoot Valley. We focused GIS mapping on specific agriculture-related attractants like livestock pasture locations, boneyards, and beehives, since these activities have impacts on grizzly bears in the landscapes we studied. We used high-quality digital aerial photographs, hydrography coverage, and detailed private landownership boundaries to focus each mapping interview on an individual ranch.

We started each mapping session by giving the rancher a 'virtual tour' of the study area by highlighting rivers and towns and then 'zooming in' on their home property. We then asked where they suspected the majority of grizzly bear activity occurred on their parcels. This type of opening put ranchers at ease and introduced them to the concept of viewing and digitizing land uses. We conducted 61 of a possible 64 mapping sessions for a response rate of 95 per cent. We had one refusal and were unable to contact the other two landowners. It was evident that, with the aid of maps, participants were able to explain their management practices readily. When we located calving or lambing areas, ranchers often told us why they had chosen specific pastures for their livestock or why they had selected a portion of their ranch for carcass disposal. For example, during the calving season, it is not uncommon for ranchers to lose calves to natural causes. Since the calving process is labour intensive and requires constant management, it was typical for boneyards to be located near calving and lambing areas, serving as a labour-saving technique for the disposal of dead animals. Mapping helped us produce spatial data sets that included the underlying explanations that ranchers helped generate and that led to other insights regarding management practices.

After we completed our GIS mapping of agricultural attractants with the ranchers, the FWP shared data on reported and verified human–grizzly bear conflict locations within our study area during the period 1986–2001. However, it took many months of cultivating the support of the FWP grizzly bear manager before we obtained the conflict database. We were then able to integrate our detailed maps of livestock management practices and beehives sites with known and verified locations of incidents involving humans and grizzly bears. This collaboration enabled us to develop empirical models that quantified patterns of conflicts and identified landscape locations that were at highest risk of experiencing conflicts (Wilson 2003; Wilson et al. 2005). We have also made our data available to the FWP.

Results

The outcomes included an improved social process, especially a better-organized arena, and better intelligence and promotional activities (Table 8.1). These process upgrades permitted the community to better address substantive matters about human-bear conflict, the key to long-term grizzly bear recovery (Primm and Wilson 2004).

The first outcome was the more effective arena, one that was better organized. This came about as we worked with a local watershed group whose activities overlapped with our study area. Throughout our time on the Rocky Mountain Front, we worked closely with the Teton River Watershed Group, a local, not-for-profit group based in Great Falls, Montana, dedicated to improving the quality and quantity water in the Teton River. The group, comprised of ranchers, landowners, and several local, state, and federal land agencies, is supportive of our work. Beginning in 1999, we attended dozens of the group's monthly meetings, and eventually we earned the trust and respect of the group. We volunteered with water-quality sampling, did monitoring

TABLE 8.1 Comparison of Rocky Mountain Front and Blackfoot Valley case studies, Montana, USA.

See text and Clark 2002a for further explanation of the social and decision-making processes terms.

Contextual Features	Rocky Mountain Front (RMF) and Blackfoot Valley (BV) Cases
Ecological setting	RMF: Foothill-grassland ecotone, riparian corridors, only region in North America where grizzly bears have maintained population presence on historic prairie grassland
	BV: Forest-grassland-wetland interface, riparian corridors, increased grizzly bear activity since mid-1990s
Social setting	RMF: Rural ranching and farming community, >100 years of settlement, medium-to-large intact ranches, sparsely populated, local conservation efforts focus on traditional land uses (Teton River Watershed Group, Sun River Watershed Group)
	BV: Rural ranching community, >100 years of settlement, small-to-medium ranches, sparsely populated, second homes and out-of-state ownership. Local conservation efforts focus on traditional land uses, habitat conservation, restoration, and landscape-scale conservation (Blackfoot Challenge).
	Commonalities: The Montana Department of Fish, Wildlife and Parks (FWP) plays prominent role in both cases for bear management. Wildlife Services plays important role during livestock depredations. Private landowners (ranchers and non-agricultural residents) own and manage significant portion of bear habitat where conflicts occur. US Fish and Wildlife Services, US Forest Service, Bureau of Land Management, Montana State Department of Natural Resources and Conservation play minor consultative roles in bear and habitat management. Culture of ranching is largely characterized by independence, strong work ethic, sanctity of private property rights, concern with economics of livelihood, and limited intrusion by state or federal agency interests.

TABLE 8.1 Comparison of Rocky Mountain Front and Blackfoot Valley case studies, Montana, USA (cont'd)

Contextual Features	Rocky Mountain Front (RMF) and Blackfoot Valley (BV) Cases
Human-bear conflict	RMF: Data from 1980s–present. Conflicts peaked in 1990s and appear to have slowed in 2000s.
	BV: Data from 1995–present, one human death in 2001. Conflicts steadily increased until 2003 and then declined thereafter.
	Commonalities: A majority of conflicts in both cases resulted from unsecured human attractants (livestock, beehives, garbage). No spatial data or analysis was present to integrate land-use practices with attractants. Data were not readily available to local residents.
Status of arena	RMF: Grizzly bear conservation and management efforts are largely driven by proactive management of FWP wildlife managers to ameliorate conflicts. However, opportunities exist to upgrade communication and partnerships with existing local conservation groups and private landowners. Landscape-level information about the nature of conflicts was available on a limited basis to local residents.
	BV: Prior to our involvement, proactive efforts were led by the FWP, but the social arena was unorganized, nearly all management decisions were made by FWP, and little collaboration existed among researchers, NGOs, or local conservation groups. Existing institutional capacity of local conservation group (Blackfoot Challenge) greatly aided the formation of partnerships through communication forums to organize and collaboratively work to resolve conflicts.
	Commonalities: Prior to intervention, management was largely dominated by FWP, social arenas were unorganized, there was a lack of resources (both personnel and financial) for proactive programs.

TABLE 8.1 Comparison of Rocky Mountain Front and Blackfoot Valley case studies, Montana, USA (cont'd)

Contextual Features	Rocky Mountain Front (RMF) and Blackfoot Valley (BV) Cases
Decision process outcomes	The process for understanding and managing conflict problems was weakly developed for all decision functions, especially in a contextual sense. The recommended standards by each function were not being approximated. The process showed there was potential to intervene and upgrade decision making.
Intelligence (planning) Standards: factual, open, and comprehensive	Areas for improvment: RMF & BV: Data were needed that integrated land-use practices, attractants, and conflict data in a spatially explicit manner. Contributions: RMF & BV: Data were needed that integrated land-use practices, attractants, and conflict data in a spatially explicit manner. Outcomes: RMF: Data were made available to wildlife managers, ranchers, and local conservation groups. BV: Data were used in a new partnership under auspices of the Blackfoot Challenge to guide specific management plans. Support of FWP and ranchers was essential for data sharing and data generation. Local knowledge is valued, local ranchers are invested in project outcomes. Research projects reflect sensitivity to ranchers' exclusive interests in private property and inclusive interests of national public through prevention of conflicts and mortality.
Promotion (debate) Standards: rational, synthetic, and effective	Areas for improvement: RMF & BV: Enhanced forums for discussion and debate were needed. Contributions: RMF & BV: For RMF, we provided summaries of our data analysis to all participants and

TABLE 8.1 Comparison of Rocky Mountain Front and Blackfoot Valley case studies, Montana, USA (cont'd)

Contextual Features	Rocky Mountain Front (RMF) and Blackfoot Valley (BV) Cases
	held a focus group meeting at the end of our project to discuss findings and possible solutions. For BV, data have been guiding participatory projects on an ongoing basis. *Outcomes:* RMF & BV: For RMF, we have encouraged wildlife managers and local conservation groups to apply our findings to their ongoing efforts. We also shared our data on land-use practices with FWP. For BV, we have taken a lead role in integrating our findings with existing wildlife management programs using a collaborative approach under the auspices of the Blackfoot Challenge.
Prescription (deciding) Standards: balanced, inclusive, and future-directed	*Areas for improvement:* BV: Existing wildlife managers were understaffed and underbudgeted. *Contributions:* BV: Multiple forums were created to invest and co-generate plans with local community. Funding and personnel were acquired. *Outcomes:* BV: Plans and projects were created that represent common interest goals among participants.
Implementation (work in field) Standards: timely, non-threatening, and common interest	*Areas for improvement:* BV: Proactive efforts by local wildlife managers were existent, but funding and personal shortages made progress slow. Efforts were not systematically organized. *Contributions:* BV: Blackfoot Challenge provided institutional capacity to develop a new partnership to raise funds and strategically target on-the-ground work. Projects were designed to reduce conflicts and minimize economic impacts on rancher livelihoods.

TABLE 8.1 Comparison of Rocky Mountain Front and Blackfoot Valley case studies, Montana, USA (cont'd)

Contextual Features	Rocky Mountain Front (RMF) and Blackfoot Valley (BV) Cases
	Outcomes: BV: Boneyards, beehives, calving areas, and other attractants were managed in new ways on the ground. Partnerships leveraged funds for electric fencing for calving, lambing areas, beehives, and waste management security.
Appraisal (monitoring and evaluation) Standards: unbiased, contextual, and constructive	*Areas for improvement:* BV: Little or no monitoring was carried out to evaluate existing efforts. *Contributions:* BV: We are helping individual ranchers and the communities appraise affects on changed management; results will be made available to all involved. *Outcomes:* BV: Ongoing monitoring of efforts will occur to determine if conflicts are reduced as a result of management changes.
Termination (stopping, exiting) Standards: timely, comprehensive, and ameliorative	*Areas for improvement:* BV: Past land-use practices led to increased conflicts. *Contributions:* BV: Collaborative and proactive programs *Outcomes:* BV: Previous management practices that caused conflict were abandoned in favour of new practices that minimized and measurably reduced conflicts. Efforts will cease once conflicts and bear mortalities have been reduced to within recommended quotas.

work, and assisted with biological control of invasive plants. By 2002, we had gained sufficient trust to partner with the Teton group in a grant to enhance riparian health. The group is currently using our GIS maps of livestock management practices for fencing calving and riparian corridors and developing off-site watering using solar pumps. We tailored the grant to benefit both ranchers and grizzly bears. The group is working on projects that will help minimize frequency of contact between grizzly bears and livestock, particularly in the spring and fall. This work has helped disseminate intelligence, but has been especially helpful in the promotion function by showing that livelihood special interests do not have to be incompatible with the minimizing of conflicts with bears. Our work helped create opportunities for improved management at individual and collective levels. Our work built on existing management and gave people the ability to make decisions about minimizing conflict.

The second outcome was improved intelligence—gathering, processing, and disseminating information to give a better understanding of both the biological and social aspects of the problem. In turn, this led to better debate and discussion as people explored the sources of conflict and their options. A key finding was that livestock ranching and beekeeping along river and creek bottoms resulted in problematic configurations of attractants in specific contexts. These were distinctly clustered arrangements of multiple livestock pastures, boneyards, and beehives. We found that attractants like boneyards and beehives were present in a majority of 'conflict hotspots'. The seasonal availability of livestock (both cattle and sheep) during calving and lambing and during spring, summer, and fall was also significant, creating repeated conflicts. In other words, while isolated attractants can lead to occasional conflicts, our analysis suggested that most conflicts (75 per cent) were concentrated in areas with the greatest densities of attractants (Wilson et al. 2006). Moreover, patterns of conflict were not a product of having livestock in grizzly bear habitats. Many parts of the study area had riparian vegetation (important seasonal grizzly bear habitat) and available attractants like calving areas where conflicts would be expected to occur but few if any did. We suggest that contexts such as these did not have the necessary collection of year-round and seasonal attractants. Apparently, isolated attractants, even in a productive habitat, are less likely to be used by grizzly bears than attractants that are concentrated (Wilson et al. 2006). This information improved intelligence by clarifying the problem and helping landowners realize that they could better manage their own actions to minimize conflicts.

The third outcome was improved debate, discussion, and exploration of management options. This outcome further reduced conflicts between local beekeepers and grizzly bears. Our work also piqued the interest of non-profit foundations that wanted to fund specific projects. Our analysis provided a scientific means to target conservation dollars on non-lethal techniques that reduced grizzly bear–beehive conflicts. For example, we directed significant funds and provided mapping services to assist Defenders of Wildlife and FWP to cost-share for 11 electric fencing projects for beehives. Our approach reflected sensitivity to local context, involved local people in participatory projects, and built local social processes or capital. This latter took the form of political support to find common interest outcomes that benefited local livelihoods and reduced conflicts with grizzly bears.

BLACKFOOT VALLEY CASE

The Blackfoot Valley has experienced a dramatic increase in bear activity since the mid-1990s. There has been extensive use by grizzly bears of river and creek bottom habitats in private ownership, and riparian corridors are likely functioning as linkage areas to potential habitats to the south (Jonkel 2002). As bear activity has increased, so has human–grizzly bear conflict, including one human fatality in 2001. This unfortunate incident and the increase in grizzly bear activity on private lands stimulated concern among residents and provided to those concerned a point of entry to (a) discover shared interests among the community, wildlife managers, researchers, and local conservation groups; (b) understand trends, conditions, and projections regarding conflicts; and (c) develop partnerships to integrate knowledge, skills, and resources to proactively develop solutions (Knight and Clark 1998). We used the following methods, as well as those described previously, in our work.

Case-Specific Methods

How did we participate and work to improve the outcomes of social and decision-making processes? By the summer of 2002, we helped co-organize a succession of meetings and discussions with private ranchers, residents, the US Fish and Wildlife Service, FWP, Brown Bear Resources, the Living with Wildlife Foundation, the Great Bear Foundation, Defenders of Wildlife, Allied Waste Services, and the Blackfoot Challenge, a local conservation group. As in the Rocky Mountain Front case, we worked with the local conservation group, the Blackfoot Challenge, a landowner-driven watershed group, to address the issues. This group proved invaluable and assisted in setting up meetings and facilitating open communication among the interested parties. These early meetings allowed us to identify our shared goals and our shared perspectives about the problem.

Early on, we conducted GIS mapping workshops that captured the attention of the Blackfoot Challenge and the ranching community. We highlighted our experiences and the results from the Rocky Mountain Front, and discussed future conflicts in the Blackfoot Valley. Having the support of the local wildlife biologist was essential. Specifically, we discussed the participatory nature of our GIS mapping, how this helped document the locations of agriculture-related attractants, and how these attractant maps can be compared with the known conflict locations according to FWP. In one GIS mapping workshop, a local rancher was willing to digitally map out his calving areas and his spring, summer, and fall pastures in front of 20 ranching neighbours. This helped illustrate the non-threatening nature of the process (Cornett 1994). Additionally, local bear managers discussed preventative techniques, including the electric fencing of calving areas. Collectively, ranchers and residents showed an interest in building on the existing proactive efforts of FWP and in replicating our GIS data-collecting efforts from the Rocky Mountain Front.

We feel that the interactive mapping—that is, asking participants to digitize and describe their land uses, was a feature that stimulated people's interest. Mapping reverses the traditional flow of information in practical problem-solving, so that rather than flowing from 'the experts' to the local people, it flows from the local people to the experts. Further, our approach created the opportunity for mutual learning. In this, the researcher is an 'expert learner'. This is distinctly different from traditional approaches to land and wildlife management, where, for example, rural people have to contend with heavy-handed regulation or being told what to do (Brick and Cawley 1996).

Workshops and meetings were crucial for defining the problem and developing shared goals. In our case, there was general consensus that reducing both human–grizzly bear conflict and livestock loss to grizzly bears was a shared and unifying goal. We discussed some of the other less tangible, but no less important, outcomes earlier. Whereas local residents did not like the increase in grizzly bear activity on their land, many ranchers were willing to partner with one another to reduce conflicts. Throughout this problem definition phase, we were particularly sensitive about how goal setting occurred. Rather than focusing the discussions on grizzly bear population numbers, bear mortality risks, bear distribution, or the importance of complying with the Endangered Species Act, we worked to craft a context-specific shared goal of preventing and reducing conflicts. This approach helped transcend philosophical differences about certain issues, such as the number of bears that should be allowed to inhabit private lands. Local ranchers were willing to participate in research and conservation efforts that they believed would benefit their security and livelihood interests. Our on-the-ground efforts reflected this concern and focused on practical innovations to reduce negative economic impacts. To date, we have completed 35 GIS mapping interviews and had only two refusals. We anticipate completing an additional 8 to 10, as interest in the program has increased recently.

Results

Our collaborative work in the Blackfoot Valley has dramatically improved the management decision process (Table 8.1). The Blackfoot Challenge was instrumental in helping to organize an arena where partnerships and forums for constructive communication were developed. After demonstrating our approach several times with small groups of ranchers and individual ranch families, we organized regular meetings. The Blackfoot Challenge offered to create a ninth working committee—the 'Wildlife Committee'. Since 2003, the Wildlife Committee has raised nearly $400,000 to support efforts. Other improvements were evident.

The first improvement was better intelligence. As we collected data from ranchers on their livestock management practices, we devised a conflict-risk ranking system to identify sites for targeting conservation dollars. Mapping involved extensive outreach and time spent with ranchers on a one-on-one basis to inventory the following: calving areas, spring cow/calf pastures, fall pastures, boneyards, sheep lambing areas, bedding grounds, and pasture configurations. We have Global Positioning System (GPS) locations for all beehives and for approximately 450 residential structures within the study area; we have obtained 30-metre Landsat imagery for vegetative cover types within the study area to assist in developing a coarse description of grizzly bear habitat; and we have hydrographic layers for locating riparian corridors throughout the study area. The development of an attractant inventory has been a basic component of our efforts and ensures a systematic approach to other decision process activities.

The second improvement was the development of multiple forums for the rational debate and discussion of options. This was done concurrently with improving intelligence. Throughout the previous two years, we held dozens of meetings and workshops. For example, the Wildlife Committee has two official work groups, the Landowner Advisory Group and the Waste Management and Sanitation Work Group, that provide overall direction for the committee. The Landowner Advisory Group is made up of key opinion leaders—local ranchers and business leaders—who help guide locally driven efforts. The Waste Management and Sanitation Work Group is made up of key industry participants and technical experts who assist with attractant security projects. These forums have worked well thus far.

The third was the development of specific action plans or prescriptions that represented a working specification of the common interest. It is important to recognize that just being able to decide is an important standard of effectiveness. In our case, we decided to develop several programs (described below) that we wanted to implement and that were supported by participants in the partnership. These prescriptions have been non-threatening, participatory, and based on economic incentives that involve cost sharing with ranchers and residents.

The fourth was the implementation of conflict abatement projects. Among these was a carcass pick-up and removal program for ranchers and electric fences to non-lethally deter bears from livestock and beehives. A citizen-based grizzly bear monitoring program was also implemented. These are described in more detail below in the descriptions of monitoring and appraisal outcomes.

The fifth improvement was upgraded monitoring and appraisal practices. Among the gains was the ranching community's widespread support of the GIS attractant inventory. As word spread throughout the community, additional ranchers wanted to participate. Our preliminary assessment of conflict risk priorities is a work-in-progress, although we have constructed 14,000 linear feet of electric fences for seven calving areas on five different ranches that were priority sites. This speaks well for the overall approach and effort.

The first official year of our carcass pick-up program ran from 20 February to 15 May 2004 to coincide with the calving season, when death losses are highest. Dead livestock act as a strong attractant to scavenging grizzly bears and can bring bears into close proximity with humans, facilities, and vulnerable livestock, especially in the early spring. The purpose of this effort is to remove carcasses from ranches so that scavenging grizzly bears do not find food rewards near human activities or have subsequent conflicts with livestock. We collected 204 carcasses in 2004, compared to 64 in 2003 for the same period, and 64 per cent of the ranchers participated in 2004, compared to 30 per cent in 2003. We collected 340 carcasses in 2005 and 306 in 2006 (Figure 8.2). Approximately 71 per cent of the active ranchers in the core project area participated in the program in 2006.

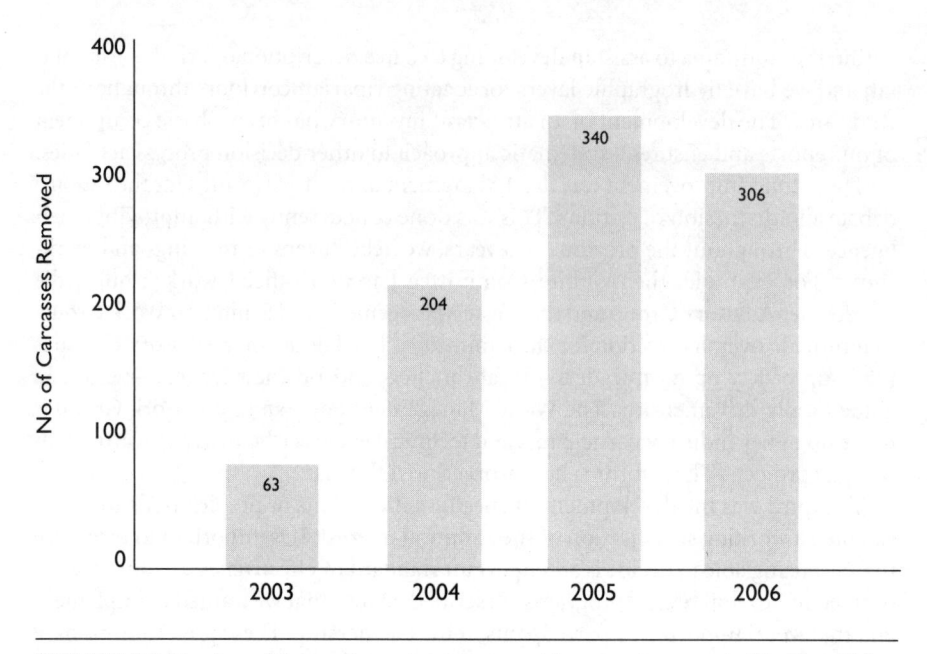

FIGURE 8.2 Number of livestock carcasses removed from project area in Blackfoot Valley, Montana, during 20 February–15 May 2003–2006

Bar chart data: 2003 = 63; 2004 = 204; 2005 = 340; 2006 = 306. Y-axis: No. of Carcasses Removed.

In the first year of the program, ranchers were reluctant to have others know the exact number of their livestock losses for fear of being perceived as deficient in animal husbandry. To overcome this obstacle, we discussed the problem with the ranchers and agreed to have a number of centralized locations where several ranchers could deposit their calves and thus preserve their anonymity during pick-up. With the support of local ranchers, many of whom are key opinion leaders in the area, we anticipate increased awareness and participation in 2007.

With respect to electric fencing, we have helped fund, design, and construct seven of the largest calving-area fences in the region. Recently, the Natural Resources Conservation Service (NRCS) started a new 'predator deterrent' program under its Environmental Quality Incentives Program to provide cost-sharing funding for projects like electric fences. As of 2006, 20 new electric fence projects for calving areas and beehives are in progress. We have helped local beekeepers prioritize hive sites for protection, and we have provided funds to construct electric fences around beehives, building on the efforts of the FWP and the Defenders of Wildlife. To date, more than 90 per cent of all beehives are protected with electric fences in the Blackfoot Valley. Our fourth program involves the monitoring of grizzly bear activity by citizens—we call this a 'neighbour network'.

The goal of the neighbour network is to have the kind of effective communication among neighbours and with FWP that would prevent human-bear conflicts from starting in the first place. We helped organize nine networks, each with a local volunteer coordinator, to facilitate communication. At present, there are nine networks with nearly 90 individual participating households that work to (a) minimize the availability of human-related attractants; (b) communicate with neighbours about bear activity; (c) report to the local coordinator any observations, incidents, and bear behaviour that may indicate problems; and (d) use aversive conditioning on grizzly bears to teach them to avoid humans and areas inhabited by humans.

Preliminary results are encouraging. Since 2003, there has been a downward trend in reported and verified human–grizzly bear conflicts in our project area. There were 77 known conflicts in 2003, 19 in 2004, and 12 in 2005. In 2003 there were 5 grizzly mortalities in the watershed, and in 2004 there were 3 known mortalities, one resulting from a hunter-related incident, one from an illegal kill, and one from a vehicle collision. In 2005 there was one road-kill mortality of a sub-adult grizzly of unknown sex. In the same year, no grizzly bears were trapped in the Blackfoot for conflict management purposes (Jonkel 2006). Recent population analysis by the US Geological Survey, which used DNA hair-snare methods, reported approximately 29 individual grizzlies in the Blackfoot area in 2004. We recognize that the downward trend in conflicts may be a result of there being fewer 'problem' bears, changes in bear foraging behaviours, or the heightened awareness of local residents, who may report conflicts less often. However, the combination of targeted prevention efforts based on our GIS analysis—the electric fencing, carcass removal, improved decision making, and citizen-based monitoring—are, in part, helping to reduce human–grizzly bear conflict. This suggests that we have helped to terminate or end harmful practices and have helped promote the use of new beneficial practices. The ranching community has stopped dumping carcasses in boneyards and has protected calves with electric fences in high-conflict areas. Also, local beekeepers have shifted practices to protect beehives. And local residents now take more active roles in monitoring bear activity. Finally, we will monitor the outcomes of our efforts until the conflicts are reduced and mortalities are down to recommended quotas.

Net Assessment from Both Cases

Other criteria have helped us evaluate the overall effort (Clark 2002a, 74–6). First, we arranged for common interests to prevail over special interests. Second, we gave precedence to high-priority rather than low-priority common interests. Third, we worked to create a process to protect both inclusive (overall well-being) and exclusive (private property rights) common interests. Fourth, where exclusive common interests conflicted, we sought to give preference to the participants whose value positions were most substantially involved; this meant that we focused on ranchers who bore the brunt of grizzly bear conflict. Fifth, we sought to allocate values (e.g., money, respect, and skill) in new ways commensurate with the nature of the problems and their solutions. This allowed ranchers to manage conflict more effectively. Finally, we sought to create a problem-solving process known for integrity and honesty.

How these additional standards played out in the two cases is evident from the case descriptions above. Specifically, our 'prototypical' participation on the Rocky Mountain Front brought needed knowledge or intelligence to help the concerned parties better understand the factors that contributed to conflicts on private agricultural lands. Throughout our work, we engaged with those parties who had the greatest influence on conflicts. We integrated key data on livestock management practices from wildlife managers and ranchers. We collaborated with local conservation groups throughout the research phases and in tangible on-the-ground projects. A replication of this effort in the Blackfoot Valley improved the outcomes of the social and decision-making processes. This allowed all participants to develop shared perceptions of the problem of human–grizzly bear conflict, shared goals, and a plan of action. We sought to be responsive to human safety and livelihood interests, rather than being stymied by ideological divides about bear numbers and whether or not grizzly bears should be allowed to inhabit private lands. The members of the local community may not like having grizzly bears on their private lands, but they recognized the value of working collaboratively (a shared interest). This shared perspective was key to all gains that were made.

LESSONS FOR INTEGRATED PROBLEM-SOLVING

The lessons described below apply to much more than the grizzly bear problem in the American West (Table 8.1). Efforts to integrate theory and practice and to move from contemplation to application in the common interest require appropriate concepts and methods, as well as settings that permit them to be used to maximum benefit. They also require that people knowledgeable and skilled in their use be central to conservation efforts. Having and using explicit standards, as discussed in this chapter, grounded our efforts and provided direction for change. Several additional lessons are evident. We will discuss them in general terms so that their universality is evident (Clark 1993).

First, improving conservation in natural resources means upgrading resource management and the policy process (not limiting oneself to getting more biological facts). In practice, this entails enhancing the procedural rationality at the heart of all efforts. Knowing how to do this is key to improving conservation. The operations that comprise procedural rationality have been known for decades and should be used in each and every case to the maximum extent possible. Problem-solving is often about getting as much rationality out of people as they are capable of giving. In fact, integrated problem-solving is really about the application of rationality in complex socio-political contexts. Appreciating embedded values and policy preferences is part of contextual problem-solving. Working successfully with these values and preferences is the challenge.

Second, problem-solvers need to be problem oriented, be aware of context, and use multiple methods, as described above. The problem-solving operations of goal clarification, trend and condition mapping, creation of future scenarios, problem definition, and the invention, evaluation, and selection of solutions are critical. Focusing on these interactive tasks in any problem setting and using multiple methods to get empirical data are essential for success. In contrast, it is all too common for natural resource professionals to construe a problem narrowly, emphasizing only a few of the above tasks and paying insufficient attention to others. For example, they often put emphasis on only the biological aspects of the problem (trends and conditions), leaving out other tasks (goals, projections, and creating alternatives) and neglecting contextual variables. Carroll et al. (2001, 978) provide an example: 'Biological science must be at the heart of any strategy to conserve carnivores, but social science, economics, law, education, and many other disciplines must be involved in the process of finding politically acceptable solutions. The proximal threats to most species are related to habitat, but the ultimate threats are human population, behaviors, and attitudes.' The irony in this statement is striking. If the 'ultimate' threat to carnivore conservation is human, then basing the 'heart' of a program on biology seems likely to fail. Approaches following that formula lead to suboptimal outcomes, weak programs, and failed efforts.

Third, we used prototyping, organizational management, and active learning as the principal means of participating and contributing to existing programs. *Prototyping* is an innovative approach geared towards learning and developing a model on which to base future actions or programs. *Organizational management* is about enhancing information processing through teams and community-based groups. *Active learning* is about both simple feedback and changing people's behaviour with respect to complex organizational and policy learning. We used these means to help people organize an arena in which conflicts could be addressed more effectively. This required our active involvement as researchers, observers, and participants. In the end, we helped people improve the outcomes of the social and decision-making processes. This entailed developing a deep understanding of the context and discovering which decision functions were working well and which were not (Table 8.1). Management improvement means intervening to upgrade the decision process in socially acceptable ways in the common interest.

CONCLUSIONS

We used prototyping successfully for two case studies in the Rocky Mountains of Montana to reduce human-wildlife conflict in the common interest. We used concepts and an analytic framework to guide our integrated research and our applied conservation efforts. Our goal was to minimize human–grizzly bear conflict through prevention. This required us to focus on changing people's perspectives and practices. This approach was necessitated by and reflects the context in both local and national conditions. We focused on the different ways people organized themselves, and we attempted to create an inclusive and effective decision-making process. To do this, we participated in a non-threatening way by leveraging collective knowledge, skills, and finances to solve problems that had relevance for the community.

To date, this approach has enhanced communication, cooperation, and problem comprehension, and has opened opportunities for problem-solving. It also produced tangible, on-the-ground outcomes that have helped resolve the problem of human–grizzly bear conflict on private lands in the American West. In the end, the solution to complex natural resource problems, such as human-wildlife conflict, lies in a problem-oriented, contextual, and multi-method approach that permits and encourages integration in all senses of the concept.

ACKNOWLEDGMENTS

We would like to thank the livestock producers and residents of the Teton watershed and Blackfoot Valley for their time, support, and involvement with this work. Very special thanks to M. Madel, J. Jonkel, and M. Long of the Montana Department of Fish, Wildlife and Parks for their vision and collaboration. We would also like to acknowledge A. and S. Rollo, C. Crawford, the Teton River Watershed Group, and the Nature Conservancy for all their help over the years with many aspects of this project. Special thanks to the Blackfoot Challenge and all members of the Wildlife Committee, the Landowner Advisory Group, and the Waste Management and Sanitation Work Group. Special thanks as well to W. Slaght, K. Kovatch, R. Burchenal, D. Iverson, R. Cahoon, L. Flemming, B. Mannix, B. Rowland, R. Hall, J. Mulcare, C. Bauer, V. Edwards, P. Sowka, M. Johnson, T. Radandt, A. Klinefelter, T. Bernd-Cohen, J. Stone, A. Vanderheiden, Dave Smith, Dave White, Glen Green, John Bowe, Cindy Frazer, Brad Weltzein, and G. Neudecker. We are grateful for financial support from the Blackfoot Challenge, the University of Montana's College of Forestry, NSF EPSCOR, the Bolle Center for People and Forests, the Wilburforce Foundation, the Brainerd Foundation, the Pumpkin Hill Foundation, the National Fish and Wildlife Foundation, the Natural Resources Conservation Service, the LaSalle Adams Fund, the Bunting Family Foundation, the Northern Rockies Conservation Cooperative, Keystone Conservation, the Montana Department of Fish, Wildlife and Parks, and the US Fish and Wildlife Service (Private Stewardship Program). Many thanks to D. Mattson, T. Merrill, R. Barrett, J. Ellis, M. Wilson, P. Wilson, and M. Patterson, who provided valuable editorial remarks on early renditions and drafts of this manuscript.

REFERENCES

Aune, K., and W. Kasworm. 1989. *Final report: East Front grizzly bear studies.* Helena, MT: Montana Department of Fish, Wildlife and Parks.

Berkes, F., C. Folke, and J. Colding (eds). 1998. *Linking social and ecological systems: Management practices and social mechanisms for building resilience.* Cambridge, UK: Cambridge University Press.

Brick, P.D., and R.M. Cawley. 1996. *A wolf in the garden: The lands rights movement and the new environmental debate.* London and Maryland: Rowman and Littlefield.

Brunner, R.D. 1991. The policy movement as a policy problem. *Policy Sciences* 24:65–98.

Brunner, R.D., C.H. Colburn, C.M. Cromley, R.A. Klein, and E.A. Olson (eds). 2002. *Finding common ground: Governance and natural resources in the American West.* New Haven, CT: Yale University Press.

Carroll, C., R.F. Noss, and P.C. Paquet. 2001. Carnivores as focal species for conservation planning in the Rocky Mountain region. *Ecological Applications* 11 (4): 961–80.

Clark, T.W. 1993. Creating and using knowledge for species and ecosystem conservation: Science, organizations and policy. *Perspectives in Biology and Medicine* 36:497–525, plus appendices.

——— 1997. Problem definition: Analytic framework, process, and outcome. In T.W. Clark, *Averting extinction: Reconstructing endangered species recovery*, 136–62. New Haven, CT: Yale University Press.

——— 2002a. *The policy process: A practical guide for natural resource professionals.* New Haven, CT: Yale University Press.

——— 2002b. Learning as a strategy for improving endangered species conservation. Special issue of *Endangered Species Update* 19 (4): 119–24.

Clark, T.W., and R.D. Brunner. 2002. Making partnerships work in endangered species conservation: An introduction to the decision process. Special issue of *Endangered Species Update* 19 (4): 74–80.

Clark, T.W., and D.J. Mattson. 2005. Appendix: Making carnivore management programs more effective: A guide for decision making. In T.W. Clark, M.B. Rutherford, and D. Casey (eds), *Coexisting with large carnivores: Lessons from Greater Yellowstone*, 271–6. Washington, DC: Island Press.

Clark, T.W., and R.L. Wallace. 2002. The dynamics of value interactions in endangered species conservation. Special issue of *Endangered Species Update* 19 (4): 95–100.

Clark, T.W., and A.R. Willard. 2000. Analyzing natural resources policy and management. In T.W. Clark, A.R. Willard, and C.M. Cromley (eds), *Foundations of natural resources policy and management*, 32–44. New Haven: Yale University Press.

Clark, T.W., R. Crete, and J. Cada. 1989. Designing and managing successful endangered species recovery programs. *Environmental Management* 13:159–70.

Clark, T.W., R.P. Reading, and A.L. Clarke (eds). 1994. *Endangered species recovery: Finding the lessons, improving the process.* Washington, DC: Island Press.

Clark, T.W., A.P. Curlee, and R.P. Reading. 1996. Crafting effective solutions to the large carnivore conservation problem. *Conservation Biology* 10:940–8.

Clark, T.W., D.J. Mattson, R.P. Reading, and B.J. Miller. 2001. An interdisciplinary framework for effective carnivore conservation. In J. Gittleman et al. (eds), *Carnivore conservation*, 223–40. New York: Cambridge University Press.

Clark, T.W., R.P. Reading, and G.N. Backhouse. 2002a. Prototyping for successful conservation: The eastern barred bandicoot program. Special issue of *Endangered Species Update* 19 (4): 125–9.

Clark, T.W., R.P. Reading, and R.L. Wallace. 2002b. Research in endangered species conservation: An introduction to multiple methods. Special issue of *Endangered Species Update* 19 (4): 106–13.

Cornett, Z.J. 1994. GIS as a catalyst for effective public involvement in ecosystem management decision-making. In V.A. Sample (ed.), *Remote sensing and GIS in ecosystem management*, 337–45. Washington, DC: Island Press.

Dale, E.E. 1960. *The range cattle industry: Ranching on the Great Plains from 1865 to 1925.* Norman: University of Oklahoma Press.

DeWalt, B.R. 1994. Using indigenous knowledge to improve agriculture and natural resource management. *Human organization* 53:123–31.

Dood, A.R., S.J. Atkinson, and V.J. Boccadori. 2006. *Grizzly bear management plan for western Montana: Draft programmatic environmental impact statement 2006–2016.* Helena, MT: Montana Department of Fish, Wildlife and Parks.

Finkel, M. 1999. From Yellowstone to Yukon. *Audubon,* July–August, 46–53.

Jonkel, J.J. 2002. *Living with Black Bears, Grizzly Bears, and Lions Project update.* Missoula, MT: Montana Department of Fish, Wildlife and Parks.

——— 2006. *Living with Predators Project: Preliminary overview of grizzly bear management and mortality 1998–2005. Region 2 Montana Fish, Wildlife and Parks.* Missoula, MT: Montana Department of Fish, Wildlife and Parks.

Jordan, T.G. 1993. *North American cattle-ranching frontiers: Origins, diffusion, and differentiation.* Albuquerque: University of New Mexico Press.

Knight, R.L., and T.W. Clark. 1998. Boundaries between public and private lands: Defining obstacles, finding solutions. In R.L. Knight and P.B. Landres (eds), *Stewardship across boundaries,* 174–91. Washington, DC: Island Press.

Lasswell, H.D. 1971. *A pre-view of policy sciences.* New York: American Elsevier.

Mace, R.D., and J.S. Waller. 1998. Demography and population trend of grizzly bears in the Swan Mountains, Montana. *Conservation Biology* 12 (5): 1005–16.

McLaughlin, G., S.A. Primm, and M.B. Rutherford. 2005. Participatory projects for coexistence: Rebuilding civil society. In T.W. Clark, M.B. Rutherford, and D. Casey (eds), *Coexisting with large carnivores: Lessons from Greater Yellowstone,* 177–210. Washington, DC: Island Press.

McLellan, B.N., and F.W. Hovey. 2001. Natal dispersal of grizzly bears. *Canadian Journal of Zoology* 79:838–44.

Madel, M.J. 1996. *Rocky Mountain Front Grizzly Bear Management Program: Four year progress report.* Helena, MT: Montana Department of Fish, Wildlife and Parks.

Mattson, D.J., S. Herrero, R.G. Wright, and C.M. Pease. 1996. Designing and managing protected areas for grizzly bears: How much is enough? In R.G. Wright (ed.), *National parks and protected areas: Their role in environmental protection.* Cambridge, MA: Blackwell Science, 133–64.

Muth, R., and J.M. Bolland. 1983. Social context: A key to effective problem solving. *Planning & Change* 14 (4): 214–25.

Primm, S.A. 1996. A pragmatic approach to grizzly bear conservation. *Conservation Biology* 10:1026–35.

Primm, S.A., and K. Murray. 2005. Grizzly bear recovery: Living with success? In T.W. Clark, M.B. Rutherford, and D. Casey (eds), *Coexisting with large carnivores: Lessons from Greater Yellowstone,* 99–137. Washington, DC: Island Press.

Primm, S.A., and S.M. Wilson. 2004. Re-connecting grizzly bear populations: Prospects for participatory projects. *Ursus* 15 (1): 104–14.

Servheen, C. 1989. The management of the grizzly bear on private lands: Some problems and possible solutions. In M. Bromley (ed.), *Bear-people conflicts,* 195–200. Proceedings of a symposium on management strategies. Yellowknife, NWT: Department of Culture and Communications.

——— 1998. Conservation of small bear populations through strategic planning. *Ursus* 10:67–73.

Taylor, D., and T.W. Clark. 2005. Management context: People, animals, and institutions. In T.W. Clark, M.B. Rutherford, and D. Casey (eds), *Coexisting with large carnivores: Lessons from Greater Yellowstone*, 28–68. Washington, DC: Island Press.

US Fish and Wildlife Service. 2003. *Grizzly bear mortalities 1992–2003*. Data set provided by US Fish and Wildlife Service Grizzly Bear Recovery Coordinator. Missoula, MT: US Fish and Wildlife Service.

Weiss, J.A. 1989. The powers of problem definition: The case of government paperwork. *Policy Sciences* 22:92–121.

Wilson, S.M. 2003. Landscape features and attractants that predispose grizzly bears to risk of conflicts with humans: A spatial and temporal analysis on privately owned agricultural land. PhD dissertation, University of Montana, Missoula.

Wilson, S.M., M.J. Madel, D.J. Mattson, J.M. Graham, J.A. Burchfield, and J.M. Belsky. 2005. Natural landscape features, human-related attractants, and conflict hotspots: A spatial analysis of human–grizzly bear conflicts. *Ursus* 16 (1): 117–29.

Wilson, S.M., J.A. Graham, D.J. Mattson, and M.J. Madel. 2006. Landscape conditions predisposing grizzly bears to conflict on private agricultural lands in the western U.S.A. *Biological Conservation* 130:47–59.

CHAPTER 9

Collaborating Experts: Integrating Civil and Conventional Science to Inform Management of Salal (*Gaultheria shallon*)

Heidi Ballard and Louise Fortmann

INTRODUCTION

Integrated resource management is likely to involve decisions about species or interspecies interactions for which there is little or no scientific data or expertise. In the absence of expertise in the community of conventional scientists, where does one turn? In a society that privileges conventional scientific knowledge, the answer is not obvious when expertise is held by people who do not have conventional scientific credentials. However, practical grounds for action lie in primatologist and feminist social theorist Donna Haraway's (1999) point that all knowledge is situated and partial. In other words, the questions you ask and what you see are a function of where you stand. Her point has been demonstrated in such fields as medicine (Martin 2001), archaeology (Conkey 2003), and biology (Asquith 1996; Thompson 2001; Gowaty 2003). Since you cannot stand everywhere at once, the prudent thing to do is to collaborate with people standing elsewhere, people with different knowledge and different ways of producing knowledge. Conventional science must integrate and acknowledge different actors and actions, including different ways of knowing, such as civil science.

The need for information on which to base the management of non-timber forest products (NTFPs) in the Pacific Northwest of the United States provides an excellent case in point.[1] NTFPs include edible mushrooms and berries, medicinal plants, and floral greens (shrubs and ferns used in floral arrangements). This chapter focuses on salal (*Gaultheria shallon*), a native understory shrub species used as a floral green that grows wild in managed and unmanaged forests of the Pacific Northwest (Figure 9.1). Increased harvest intensity in the last several decades has caused a marked decrease in the availability of commercial-quality salal in some areas, according to harvesters (Brown and Marin-Hernandez 2000) and managers. Because floral greens species have historically been considered weeds in forests under intensive timber management, research on them has been limited to their competition with timber species

FIGURE 9.1 Salal (*Gaultheria shallon*) and bracken fern (*Pteridium aquilinum*) form a dense understory in this intensively managed stand of Douglas fir (*Psuedotsuga menziesii*).

and response to timber management practices such as thinning, herbicides, and fertilizers (Bailey et al. 1998; Thomas et al. 1999; He and Barclay 2000). Very little scientific literature has been published on the consequences of intensive harvesting and management on the growth and reproduction of the NTFP species themselves (see Cocksedge 2003). In the case of salal harvesting, the scientific knowledge needed to combine the management of understory and overstory species was obtained by integrating the knowledge and practices of conventional and civil scientists through a kind of participatory research that recognized all parties to the research as scientists.[2] We call this practice *interdependent science*—the practice of integrating and utilizing both conventional and civil science. This chapter will use the case of the Salal Project to describe the need for interdependent science in integrated resource management, how this can be done, and why it is important.

EXPERTS AND EXPERTISE: CIVIL, CONVENTIONAL, AND INTERDEPENDENT SCIENCE

Our understanding of science comes from Donna Haraway's (1999, 182) statement that the goal of science is to create 'better accounts of the world'. These 'better accounts of the world' are then applied to resource management. With Haraway's definition as the fundamental characteristic of science, we define civil, conventional, and interdependent science as follows:

Civil science[3] is a set of knowledge-producing practices intended to provide better accounts of the world. Civil scientists often work informally, using experimental and observational techniques that they have developed themselves, and apply what they learn directly to local resource management. Their science depends on their knowledge of a particular set of social-ecological relationships. Their concrete findings, validated by utility, are often particular to a place, but at times they may be generalizable. Because civil science is likely to be a localized, individual practice, neither the civil scientists nor their findings will necessarily travel far, if at all.

Conventional science[4] is also a set of knowledge-producing practices intended to provide better accounts of the world. It contrasts with civil science in that university-educated conventional scientists almost always work in a formal organization using prescribed experimental and observational techniques. Their science depends on access to networks of conventional scientists, external funds, equipment, and legitimizing institutions such as scholarly journals. Their findings, often validated by statistical tests and networks of other scientists and journals, are intended to be generalizable. They may not translate easily into useful solutions to local management problems but may inform large-scale policy formation. Because of the social organization of conventional science, both conventional scientists and their findings travel with relative ease.

Interdependent science[5] is a set of knowledge-producing practices intended to provide better accounts of the world through collaboration between conventional and civil scientists. We wish to be clear that we are not arguing that civil science should simply serve as a new source of knowledge to inform conventional science. Rather, we mean that there should be new practices for the co-production of knowledge, that is, interdependent science. We are using the Salal Project as an example of the application of interdependent science to forest management.

The emergence and recognition of interdependent science require that we grapple with the definitions of experts and the concept of objectivity. The iconic figure of conventional science is usually the objective scientific 'expert'. However, we argue that the metric for deeming an activity scientific should be outcomes (knowledge and understanding), the processes that lead to that knowledge, and with them, better accounts of the world. Further, the objective scientific expert is more often encountered as an ideal type rather than as an individual in real-world situations.

Present-day understandings of actual scientific practice have long since moved from the image of the scientific expert working alone to one of complex social interactions. Latour (1987) describes how the creation of facts is a collective social process involving a wide variety of actors rather than a single moment of individual genius. The practice of crediting a single or small number of authors has been recognized as masking the numbers and variety of people who have contributed significantly to scientific research and writing (Boyle 1996; Haraway 1997). Further, the history of science shows that, while it is generally not acknowledged, there have been many civil science contributions to the development of conventional science for at least five centuries. Although it is linked with Linnaeus, botany had its roots in South Asian local ecological knowledge, with important contributors being women and other low-status people (Grove 1995). Nineteenth- and twentieth-century British scientific forestry (Sivaramakrishnan 1999), twentieth-century US plant breeding (Murphy 2004), and international pharmacology (http://biotech.icmb.utexas.edu/botany/perihist.html) have all incorporated the knowledge and practices of civil scientists, sometimes unknowingly, sometimes unwillingly. Just as resource management practices can be improved by integrating different disciplinary approaches to management, the science that informs it can be improved by integrating the different knowledge and experience from civil and conventional scientists. The Salal Project to which we now turn demonstrates the necessity of intentional collaboration between civil and conventional scientists, the integration of their practices and knowledge (that is, interdependent science), to produce rigorous policy and management relevant research.

THE SALAL PROJECT SETTING

In 2001 a small group of NTFP harvesters founded the Northwest Research and Harvesters Association (NRHA) in the Olympic Peninsula in an endeavour to gain access to NTFP resources and to collaboratively manage and monitor these resources, on which they depend. Later that year, the NRHA expanded when it formed a partnership with Heidi Ballard, the resulting entity called the Olympic Peninsula Salal Sustainability and Management Research Project (the Salal Project). This project's objectives were (1) to design and implement harvest impact studies on salal; (2) to document harvester local ecological knowledge;[6] and (3) to develop an effective methodology for monitoring the harvesting and condition of the resource (Ballard et al. 2002). As a result, Ballard was also able to analyse the way federal, state, and private salal management approaches affect harvest practices, assess the sustainability of the resource, and provide feedback and recommendations to managers (Ballard 2004).

A harvester is defined here as anyone who picks and sells salal commercially for any part of the year. Harvesters must buy permits or leases to gain legal access to the plants because almost none of them own their own land. Harvesters in the study area are predominantly undocumented male migrant workers from Mexico and Latin America (Hansis 2002; Lynch and McLain 2003) and immigrants from Southeast Asia. Many have been harvesting on the Olympic Peninsula for fewer than 10 years, and most speak little English. On some types of landownerships, workers with few English-speaking skills can obtain a permit with no documentation (or need not obtain a permit at all), find transportation into the woods, and sell their product each day for cash, with less threat of deportation than in many other types of work (although fear of deportation is common and constant). Most harvesters get a large portion of their income from salal harvesting and the harvesting or handling of Christmas greens (conifer boughs), but they may pick other NTFPs when their prices are higher. Despite their direct impact on and knowledge of the resource on which their livelihood depends, harvesters have little influence on official forest management policies and practices.

Harvesters making decisions about where, how much, and how to harvest floral greens on the Olympic Peninsula face a confusing array of ways to gain access to salal.[7] Land is owned by private individuals and corporations and by the state and federal government. Legal access to floral greens is a patchwork of short- and long-term permits and leases available for widely differing amounts of money, each with varying rules on who may harvest when and how. Some large-scale landowners attempt to simplify the process by giving long-term contracts to wholesale floral greens companies with sufficient capital, though not all wholesalers lease land. In order to get access to land that is controlled by a wholesaler, harvesters must join crews authorized to pick there and, in most cases, must sell their product back to that wholesaler. This generally means that if they hope to renew their permit to continue harvesting in that area, harvesters receive a lower price than they could from other wholesalers. Harvesters and others inside the industry say that this is standard practice, though most wholesale companies deny that this relationship exists. Harvesters can feasibly make a good income if they have their own transportation into the forest and can negotiate with landowners for permits and with buyers for good prices. However, many end up paying for transportation from a driver, paying an inflated price for a permit, and giving a percentage of the day's earnings to a 'padrón', the person who acts as unofficial liaison with the buyer. Given these conditions, stealing, or 'unpermitted' harvesting, occurs regularly on both private and public lands, often overwhelming any planned management practices on the part of land managers and harvesters with permits.

THE PROCESS OF INTEGRATING HARVESTER ECOLOGICAL KNOWLEDGE AND CONVENTIONAL SCIENTIFIC KNOWLEDGE

With its purpose being to inform forest management of NTFPs on the Olympic Peninsula, the Salal Project collaboration began when the president of the NRHA, Don Collins, and Heidi Ballard decided to conduct research on salal harvest sustainability. Because the study of the ecology and management of NTFPs is a relatively new pursuit for ecologists in the United States, and because harvesters could potentially contribute valuable knowledge and experience to the research, Collins and Ballard decided that an interdependent science approach would be essential to the research design process. The salal harvesters and Ballard each had unique knowledge and experience, such that none of them could conduct the research effectively without the others. Therefore, in addition to reporting the results of the ecological experiment (Ballard 2004), the Salal Project documented the ecological knowledge of the harvesters through interviews and participant observation. The harvesters who participated included women and men, Latinos, Southeast Asians, and Anglos, and ranged in harvest experience from 3 months to 25 years. The combination of harvesters' knowledge and conventional scientific knowledge produced the integrated interdependent science that made this project possible.

The framing of the research question for the Salal Project serves as an example of the interplay described above. Initially, Ballard considered focusing the research on the incentives for, and effects of, harvesting different commercial grades of salal. However, many harvesters specified that the way they harvested was a consequence of whose land they were picking and what kind of permit or access system was in place. The research was thus reframed to include harvesting intensity. When asked to describe how they pick salal, harvesters responded as follows:

- 'I will make sure my crew picks so the land looks good and healthy the next year because the owner trusts me and I want to do good for him. After my two-year lease, I will maybe try to get the same land again.'
- 'It's nice to let the plant grow if you have enough land, but usually you don't, so you have to pick pretty hard.'
- 'I know I have my lease for next year so I can leave it to grow, but other people with [short-term] permits take it all, they don't care about it growing back.'
- 'I always see in the future, if we leave something, next year we'll have it. If we cut it all, it's done; next year maybe there's nothing.'

Harvester Hypotheses

Integration of conventional and civil science began with collaborative hypothesizing about how salal responds to harvests at different intensities. Harvesters suggested that when people picked 100 per cent of the commercial-quality product in a given area (called 'heavy intensity harvest' for the experiment), it involved removing mature stems along with new leaf buds, such that if these plants were picked every year for several years, growth would decrease. They explained that the plant would not be able to maintain growth levels and therefore this harvest method would not be sustainable over time. Removing only 33 per cent of available commercial shoots every year (called 'light intensity harvest' for the experiment), harvesters hypothesized, would not decrease growth because these new leaf buds were not all removed. When Ballard suggested testing these two methods in a 'conventional' controlled experiment across different sites, several harvesters volunteered to help with the experiment.[8]

Harvesters made clear links between different harvesting practices and the sustainability of the resource and, therefore, their livelihood. This helped frame both the hypotheses and the interpretation of the results.

- 'I pick for high quality, most people pick now for volume. I won't pick it if it has any bug chew, so next year it'll be thicker and might shoot out more stems.'
- 'I cut into the green [stem] only, only two years of growth; it's wood after that. In the old wood you can see there's no buds that will grow.'
- 'When you pick higher up on the plant, the little bumps below grow to new stems next year.'
- '... the brush needs to rest ... should let an area rest after picking it for 2–3 years. We had an area ... we picked for 5 or 6 years [in a row] and the brush became less and less.'

Interdependent Experimental Design

Harvesters contributed substantially to the experimental design of the Salal Project because of their considerable ecological knowledge about the effects of stand conditions on understory species, particularly commercial-quality salal. For example, nearly all harvesters described in various terms the stand conditions, such as dominant tree species and amount of canopy closure, required to produce commercial-quality salal (also referred to as 'brush'):

- 'You can tell where it's good salal ... because there needs to be spaces for light to get to the ground.'
- 'When the trees are young, everything is growing good, then when the trees get bigger and there's more shade, there's less salal.'
- 'When you see alder, there's no brush.'
- '[I look for] lowlands, where Douglas fir grows; some lands have a lot of maple trees but salal doesn't grow under maples.'

Additionally, many harvesters described relationships among understory species, noting that salal is found in association with some other shrub species but not others. Both the canopy and understory descriptions by harvesters corroborate the conventional science definitions for the vegetation associations of the Olympic National Forest (Henderson et al. 1989). Harvesters' observations supplemented the associations' descriptions, however, by specifically targeting commercial quality.

- 'Good tall salal grows in the middle of green huck (*Vaccinium ovatum*), it grows good with huck.'

- 'It grows really well between huck, it grows longer, but then it's hard to walk through so it's harder to pick a lot fast.'

Several harvesters also described elevation, soil moisture, and other soil conditions as important for salal growth and quality. Conventional forest ecologists recognize these factors as being among those that determine understory species distribution and dominance in the region (Franklin and Dyrness 1973), but they know less about how these factors affect the distribution of commercial-quality characteristics.

- 'From the road we look for big trees and green so we know under there is a lot of wet lands, the soil is wet.'
- 'When the roots are tight and strong in the ground, [the leaves are better].'
- 'Altitude of [about] 2800 ft . . . if it's the right altitude, you find salal.'

By integrating the harvesters' knowledge of overstory, understory, and environmental characteristics that affect salal growth and commercial quality, the experimental design that had relied solely on a review of the literature and discussions with ecologists was significantly changed. Research sites chosen by harvesters varied by elevation and forest stand type. Most important, the harvesters and Ballard collaborated to identify and operationalize variables not commonly found in the conventional scientific literature, creating an integrated protocol that would specifically measure impacts of the harvesting on the plant. In the summers of 2001–3 harvesters worked as field assistants, collecting and recording data (Figure 9.2). In addition, four US Forest Service technicians also collected data, often in teams that included salal harvester field assistants. Complementary to the harvesters' civil science contributions were the statistically sound experimental design and plant ecology field methods that Ballard's conventional science background provided.

Had ecologists not collaborated with harvesters, it would have been of limited practical value for them to harvest salal experimentally in the standardized ways of the discipline; the results of such an experiment would not have reflected actual harvest situations and would not be applicable to impacts that landowners and harvesters actually face. Indeed, at the start of the Salal Project, several ecologists suggested that such harvest experiments should occur in a greenhouse under controlled conditions. While this would certainly have been an effective way to reduce variation, control environmental and anthropogenic variables, and measure accurately the effects of biomass removal on salal regrowth, it would ignore the real harvest practices used by harvesters across the Olympic Peninsula. Instead, harvesters helped design the research to test the light and heavy intensity harvest practices *in situ* on public and private lands, where harvesting actually occurs. Hence, some experimental control was sacrificed for better accuracy of harvest conditions.

Defining Harvest Treatments That Reflect Tenure Conditions

Collaboratively defining the harvest treatments on the basis of the relationship between harvest practices and tenure resulted in an experiment that reflected real-world practices but also relied on standardized ecological guidelines for biomass

FIGURE 9.2 A salal harvester scientist records information on data sheets translated into Spanish.

removal experiments. In the fall of 2001 and 2002, harvesters applied the harvest treatments, which included weighing and taking samples of the product harvested. The precise definition of *available commercial shoots* had to accurately represent true harvest practices, while the quantifying of the amount of biomass removed had to satisfy the standards of consistency in ecological research. So that both would be achieved, a core group of harvesters and the university researcher collaboratively determined the harvest treatments and applied them over the three harvest seasons of the study. This integration of conventional and civil science knowledge reflects the interdependent science approach. Methods from the literature had to be modified to encompass the parts of the plant actually harvested. Harvesters had to compromise efficiency of harvesting for thoroughness and uniformity of biomass removal. Rather than representing a loss of validity, this compromise represented the successful navigation of two sets of standards, resulting in a more valid experimental design. Further, because the treatments were defined as an outcome of specific access conditions for harvesters, the experiment had a direct link to policy.

Interpreting the Results Together

Even though Ballard did the statistical analysis of the harvest yield data of the experiment, harvester knowledge and experience played a crucial role in the data analysis. The experimental results were initially unexpected but became understandable when harvesters described their particular observations about patterns of insect herbivory and fungal disease spread. Most harvesters described insect damage and several different fungal diseases that afflict the plant, offering hypotheses on what conditions cause the spread of different diseases. Since the commercial value of salal depends on leaf quality, any disease and other blemishes on the leaves are of great concern to most harvesters. The harvesters' observations played an important role in the choice of the variables to be measured during the experiment, as well as in the interpretation of results that seemed inconsistent without consideration given to insect and disease spread. The following are some of the observations of the harvesters:

- 'Some areas have a lot of bugs, they chew holes in the leaves and you can't pick there.'
- 'You have to pick early in the season because the longer you wait, the more bug chew there is.'
- 'You have to be careful about the brown spot, because if you pack it with your other bunches, it spreads through the whole bundle.'

In September of 2003, 20 harvesters in the NRHA gathered to interpret the harvest yield results for each year for each experimental site using large bar graph representations of the results. After being instructed on how to read bar graphs and with several harvesters serving as Spanish translators and facilitators for small-group discussions, the harvesters discussed why some results differed from their hypotheses, why sites responded differently to the same harvest treatments, and how the results could be used for management recommendations. Several harvesters pointed out how environmental conditions might have affected the results, one man saying, 'You know how it was a really dry year last year, and so some areas had more bugs [eating the leaves] than others. Maybe we got different [yields] for different areas because the bugs ruined [the quality] of some areas.'

It is important to note that despite the importance of these interpretations to the research project, most harvesters who participated in the analysis later explained that they did not consider that meeting to have been part of the scientific research process, but rather to have been just another social dinner gathering.

Questions of academic rigour and validity are regularly raised in debates on the potential success or failure of integrating science with a participatory research approach (Bradbury and Reason 2003). Depending on the types of questions being asked and on whether funding was available, a combination of both highly controlled experiments and harvester-designed experiments could be employed to take advantage of the benefits of the two types of research for the purpose of answering management questions. In the case of the Salal Project's integrated, interdependent science, each step of the research process involving harvesters required negotiations about the accepted level of rigour for ecological research, the principles of participatory research and integrating civil and conventional science, and the goal of policy and management relevance.

CONTRIBUTIONS TO MANAGEMENT AND POLICY

Harvesters' knowledge and experience were also specifically important with respect to management recommendations made to local forest managers (Ballard 2004). In addition, the harvesters pointed out aspects of forest management affecting salal harvest and production that were not tested in the experiment. Several harvesters explained why salal responded positively to the thinning of trees, a phenomenon well known to foresters. Thinning opens up the canopy and allows light to penetrate to the forest understory, stimulating growth of the salal and other floral green species.

- 'Where they've thinned—3 or 4 or more years later, it's good thick brush.'
- 'Some companies do a 50% timber cut—selective (cut)—the sun hits the ground more, it kills the big brush, but 2–3 yrs later the brush is 3 times more production, brush is so thick. The sun and the logging activity make it grow more; also the planting makes it grow more. Compared to a clearcut, [when they] take 100% of the trees, it takes 10–15 years to get good brush, when trees [are] tall enough.'

Harvesters referred to a variety of other silvicultural activities, some describing their observed effect on salal, others explaining their interpretations of the cause of those effects. Some of them described effects of timber management practices (fertilizing, for example) that are only recently being explored by conventional forest researchers. Though the effects of silvicultural practices on commercial salal production were not tested in this experiment, harvesters' observations about these effects were integrated into management recommendations for the joint production of timber and non-timber products for state and federal land managers (Ballard 2004).

- 'Sometimes when (the timber company) sprays fertilizer, the salal uses it first.'
- 'When the trees are planted close together, salal doesn't grow that well. But when trees are bigger and more separated, the salal is better, maybe because the trees are bigger and hold the moisture better.'

The fact that harvesters' knowledge, observations, and recommendations were explicitly integrated into the recommendations made to local forest managers is significant. By involving harvesters in the analysis of the results, Ballard hoped to 'broaden the bandwidth' of questions of validity and rigour in research (Bradbury and Reason 2003) and thereby challenge managers and scientists to expand their definitions of useful knowledge for research and practice. Specifically, harvesters' observations and recommendations about the sustainability of rest-rotation management of salal harvest areas were supported by the experimental results and were offered to forest managers as a key management (and 'permitting') recommendation of the Salal Project. Although ecological outcomes will take years to surface, the interdependent science approach to the study of salal harvest impacts will potentially result in positive outcomes both for policies of NTFP harvest permitting and access and for the management of NTFP's by public land management agencies on the Olympic Peninsula.

The results of the sustainability study were presented to the Forest Service district and forest supervisors, the Washington Department of Natural Resources land managers and foresters, and private timber company personnel. Because official forest managers lack information on salal harvesting and its impacts, both the ecological and the permitting recommendations can be integrated into better forest management practices and can simultaneously improve harvester livelihoods in the region. For example, public and private land managers were surprised by the harvesters' extensive knowledge of timber management practices and how they affect understory species. In many cases, the land managers had assumed that harvesters either did not know about sustainable management practices and forest ecology or did not care, a misconception that resulted in the exclusion of harvesters from participation in management or research. As the research and management activities of the NRHA have become more well known, several private and public land managers changed their attitude towards the harvester organization and began negotiations about exclusive resource access for harvesters in exchange for the harvesters' monitoring and research.

Several harvesters who regularly use short-term (two-week) permits explained that they know the way they harvest 'hurts the plant' but that if they had an exclusive lease of their own for several years, they would use a less-intensive harvest method. Because the harvest treatments were designed to reflect decisions made by harvesters under certain existing access and permitting conditions, the research results can inform and possibly influence those conditions. The study suggests that harvesters use light harvest practices when they have long-term access to the land and heavy harvest practices when they are harvesting in a short-term or open-access system. The experimental results suggest that the heavy harvest treatment might have stimulated growth in the short-term, seemingly contradicting the original hypotheses. However, the experiment only occurred over a short period of two years, and further research is needed to determine longer-term effects. The information resulting from the integration of civil and conventional science (interdependent science) will directly contribute to informing permitting policies that can improve harvester livelihoods, potentially increasing their input into forest management and their ability to sustainably manage the resource.

The results of the Salal Project have integrated-management implications for silvicultural practices on public lands. Federal and state public land managers have requested that the experimental results of the Salal Project be used in forest management decision making. The research will provide new information on relationships between overstory characteristics and salal commercial quality, largely owing to the knowledge and experience that harvesters contributed to the research process. This information will allow public land managers to consider the overall economic and environmental costs and benefits to timber and non-timber product species of various silvicultural treatments, such as thinning, pruning, and clear-cutting. Further, this concrete example of the process and outcomes of integrating conventional and civil science may make it easier for forest managers to expand the sources of knowledge on which they draw when making decisions.

CONCLUSIONS

The case of floral greens demonstrates that our very best conventional science has not been sufficient to address questions of forest ecology with pressing management implications. If we were to assume that our very best conventional scientific knowledge is *not yet* sufficient, our quick policy solution might be to improve conventional science with additional funding. But if we assume, on the basis of available data and theories, that our very best conventional scientific knowledge, good as it is, is *unlikely* to be sufficient to address our problem, our policy might then be to integrate our best conventional science with other ways of producing knowledge, in this case knowledge gained through on-the-ground practice and observation of harvesting. The Salal Project serves as an example of how a marginalized group of civil scientists, not traditionally considered part of the local community, can be important knowledge producers. Their knowledge, combined with conventional ecological knowledge, can produce interdependent ecological research and improved natural resource management practices and policies.

The Salal Project also illuminates how knowledge and expertise are developed in different ways. Local technical knowledge in this case was developed by harvesters working in, indeed depending on, a local ecological system. Salal harvesters came into a new ecosystem and, like all scientists, observed, experimented, and learned. Without their practice of civil science, the conventional scientist, Ballard, would not have been able to do rigorous, policy-relevant research.

We have seen that expertise can have many sources and that 'experts' can wear many hats, each expert having the ability to contribute to rigorous research. And we have seen that creating better accounts of the world, which is the goal of science, requires recognizing when different kinds of knowledge and knowledge production are essential for, and can be integrated to produce, successful and relevant research and management.

NOTES

1. NTFPs, which have become an industry worth more than $100 million and employing over 10,000 people (Schlosser et al. 1991) in this region, are ecologically important for the habitat and nutrients they supply to a variety of plants and animals (Molina et al. 1998). Joint production of NTFPs and timber in this region may provide the economic incentives for forest managers to adopt management strategies that conserve biodiversity and provide other ecological benefits (Alexander et al. 2002, Kerns et al. 2003).
2. Information on participatory methods can be found in Ashby and Sperling 1995; Chambers 1994; Greenwood et al. 1993; Minkler and Wallerstein 2003; Park 1997; Pattengill-Semmens and Semmens 2003; Usher 2000; and Western and Wright 1994.
3. We use the term *civil science* in reference to civil society; however, this term is often interchangeable with *civic, citizen, participatory*, and *lay science*. For more information, see

Bäckstrand 2003; Bagby and Kusel 2003; Kruger and Shannon 2000; and Reed and McIlveen 2006.
4. We use the term *conventional science* rather than the often-used term *Western science* because of the false implication that all knowledge systems in the Western countries of North America and Europe are based on Newtonian science and that those in non-Western countries are not. See Berkes et al. 2000.
5. The term was first used to our knowledge by Murphree (2004).
6. Additional information on the role of local ecological knowledge can be found in Berkes and Folke 1998; Mackinson 2001; Mallory et al. 2003; Nabhan 2000; Nyhus et al. 2003; Olsson and Folke 2001; Stevenson 1996; and Ticktin and Johns 2002.
7. Information on resource access and tenure can be found in Bruce et al. 1993; Ostrom 1990; Peluso 1996; and Ribot and Peluso 2003.
8. Though harvesters volunteered their time for the first year, in subsequent years funding was obtained to pay harvesters for their time working on the experiment.

REFERENCES

Alexander, S., D. Pilz, N.S. Weber, E. Brown, and V.A. Rockwell. 2002. Mushrooms, trees, and money: Value estimates of commercial mushrooms and timber in the Pacific Northwest. *Environmental Management* 30 (1): 129–41.

Ashby, J.A., and L. Sperling. 1995. Institutionalizing participatory, client-driven research and technology development in agriculture. *Development and Change* 26:753–70.

Asquith, P. 1996. Japanese science and Western hegemonies: Primatology and the limits set to questions. In L. Nader (ed.), *Naked science: Anthropological inquiry into boundaries, power and knowledge*, 239–56. New York: Routledge.

Bäckstrand, K. 2003. Civic science for sustainability: Reframing the role of scientific experts, policy-makers and citizens in environmental governance. *Global Environmental Politics* 3 (4): 24–41.

Bagby, K., and J. Kusel. 2003. *Civic science partnerships in community forestry: Building capacity for participation among underserved communities*. Taylorsville, CA: Pacific West Community Forestry Center.

Bailey, J.D., C. Mayrsohn, P.S. Doescher, E. St Pierre, and J.C. Tappeiner. 1998. Understory vegetation in old and young Douglas-fir forests of western Oregon. *Forest Ecology and Management* 112 (2): 289–302.

Ballard, H. 2004. *Harvester knowledge and science: Participatory research on the impacts of harvesting salal (Gaultheria shallon) on the Olympic Peninsula, Washington*. PhD dissertation, University of California, Berkeley.

Ballard, H., D. Collins, A. Lopez, and J. Freed. 2002. Harvesting floral greens in western Washington as value-addition: Labor issues and globalization. Paper presented at meeting of the International Association for the Study of Common Property, Victoria Falls, Zimbabwe. http://dlc.dlib.indiana.edu/archive/00001077/.

Berkes, F., J. Colding, and C. Folke. 2000. Rediscovery of traditional ecological knowledge as adaptive management. *Ecological Applications* 10 (5): 1251–62.

Berkes, F., and C. Folke (eds). 1998. *Linking social and ecological systems: Management practices and social mechanisms for building resilience*. Cambridge, UK: Cambridge University Press.

Boyle, J. 1996. *Shamans, software, & spleens: Law and the construction of the information society.* Cambridge, MA: Harvard University Press.

Bradbury, H., and P. Reason. 2003. Issues and choice points for improving the quality of action research. In M. Minkler and N. Wallerstein (eds), *Community-based participatory research for health.* San Francisco, CA: Jossey-Bass Inc.

Brown, B.A., and A. Marin-Hernandez. 2000. *Voices from the woods: Lives and experiences of nontimber forest workers.* Wolf Creek, CA: Jefferson Center for Education and Research.

Bruce, J., L. Fortmann, and C. Nhira. 1993. Tenures in transition, tenures in conflict: Examples from the Zimbabwe social forest. *Rural Sociology* 58 (4): 626–42.

Chambers, R. 1994. The origins and practice of participatory rural appraisal. *World Development* 4 (7): 953–69.

Cocksedge, W. 2003. Ecological and social aspect of the NTFP, salal (*G. shallon*) in Vancouver Island. Master's thesis, University of Victoria.

Conkey, M.W. 2003. Has feminism changed archeology. *Signs* 28 (3): 867–80.

Franklin, J.F., and C.T. Dyrness. 1973. Natural vegetation of Oregon and Washington. *US Forest Service General Technical Report PNW* 8:1–417.

Gowaty, P.A. 2003. How feminism changed evolutionary biology. *Signs* 28 (3): 901–21.

Greenwood, D.J., W. Foote Whyte, and I. Harkavy. 1993. Participatory research as a process and as a goal. *Human Relations* 46 (2): 175–92.

Grove, R.H. 1995. *Green imperialism: Colonial expansion, tropical island Edens, and the origins of environmentalism, 1600–1860.* Cambridge, UK: Cambridge University Press.

Hansis, R. 2002. Case study, workers in the woods: Confronting rapid change. In E.J. Jones, R.J. McLain, and J. Weigand (eds). *Nontimber forest products in the United States.* Kansas: University of Kansas Press.

Haraway, D. 1997. *Modest_Witness@Second_Millennium.FemaleMan_Meets_OncoMouse™: Feminism and technoscience.* New York: Routledge

——— 1999. Situated knowledges: The science question in feminism and privilege of partial perspective. In M. Biaglioli (ed.), *The science studies reader,* 172–88. [1988] New York: Routledge.

He, F., and H.J. Barclay. 2000. Long-term response of understory plant species to thinning and fertilization in a Douglas-fir plantation on southern Vancouver Island, British Columbia. *Canadian Journal of Forestry* 30:566–72.

Henderson, J.A., D.H. Peter, R.D. Lesher, and D.C. Shaw. 1989. *Forested plant associations of the Olympic National Forest.* USDA Forest Service Pacific Northwest Region R6-ECOL-TP 001-88.

Kerns, B.K., D. Pilz, H. Ballard, and S. Alexander. 2003. Managing for nontimber forest resources. In A.C. Johnson, W. Haynes, and R.A. Monserud (eds), *Compatible forest management: Case studies from Alaska and the Pacific Northwest.* Kluwer Academic Publishers.

Kruger, L., and M. Shannon. 2000. Getting to know ourselves and our places through participation in civic social assessment. *Society and Natural Resources* 13:461–78.

Latour, B.1987. *Science in action: How to follow scientists and engineers through society.* Cambridge, MA: Harvard University Press.

Lynch, K., and R.J. McLain. 2003. Access, labor, and wild floral greens management in western Washington's forests. General technical report, Pacific Northwest Research Station, USDA Forest Service, July 2003, 1–61.

Mackinson, S. 2001. Integrating local and scientific knowledge: An example in fisheries science. *Environmental Management* 27 (4): 533–45.

Macpherson, C.B. 1978. *Property: Mainstream and critical positions.* Toronto: University of Toronto Press.

Mallory, M.L., H.G. Gilchrist, et al. 2003. Local ecological knowledge of ivory gull declines in Arctic Canada. *Arctic* 56 (3): 293–8.

Martin, E. 2001. *The woman in the body: A cultural analysis of reproduction.* Boston: Beacon Press.

Minkler, M., and N. Wallerstein. 2003. *Community-based participatory research for health.* San Francisco, CA: Jossey-Bass Inc.

Molina, R., N. Vance, et al. 1998. Special forest products: Integrating social, economic, and biological considerations into ecosystem management. In K.A. Kohm and J.F. Franklin (eds), *Creating a forestry for the 21st century: The science of ecosystem management.* Washington, DC: Island Press.

Murphree, M.W. 2004. Communal approaches to natural resource management in Africa: From whence and to where? Keynote address, Breslauer Symposium, 5 March 2004, University of California at Berkeley.

Murphy, R.P. 2004. Personal communication.

Nabhan, G.P. 2000. Interspecific relationships affecting endangered species recognized by O'odham and Comcaac cultures. *Ecological Applications* 10 (5): 1288–95.

Nyhus, P.J., Sumianto, et al. 2003. Wildlife knowledge among migrants in southern Sumatra, Indonesia: Implications for conservation. *Environmental Conservation* 30 (2): 192–9.

Olsson, P., and C. Folke. 2001. Local ecological knowledge and institutional dynamics for ecosystem management: A study of Lake Racken Watershed, Sweden. *Ecosystems* 4 (2): 85–104.

Ostrom, E. 1990. *Governing the commons: The evolution of institutions for collective action.* Cambridge, UK: Cambridge University Press.

Park, P. 1997. Participatory research, democracy and community. *Practicing Anthropology* 19 (3): 8–13.

Pattengill-Semmens, C.V., and B.X. Semmens. 2003. Conservation and management applications of the reef volunteer fish monitoring program. *Environmental Monitoring and Assessment* 81 (1–3): 43–50.

Peluso, N.L. 1996. Fruit trees and family trees in an anthropogenic forest: Ethics of access, property zones, and environmental change in Indonesia. *Comparative Studies in Society and History* 38 (3): 510–48.

Reed, M., and K. McIlveen. 2006. Toward a pluralistic civic science? Assessing community forestry. *Society and Natural Resources* 19:591–607.

Ribot, J., and N.L. Peluso. 2003. A theory of access. *Rural Sociology* 68 (2): 153–81.

Schlosser, W., K. Blatner, et al. 1991. Economic and marketing implications of special forest products harvest in the coastal Pacific Northwest. *Western Journal of Applied Forestry* 6 (3): 67–72.

Sivaramakrishnan, K. 1999. *Modern forests: Statemaking and environmental change in colonial Eastern India.* Stanford, CA: Stanford University Press.

Stevenson, M.G. 1996. Indigenous knowledge in environmental assessment. *Arctic* 49 (3): 278–91.

Thomas, S.C., C.B. Halpern, D.A. Falk, D.A. Liguori, and K.A. Austin. 1999. Plant diversity in managed forests: Understory responses to thinning and fertilization. *Ecological Applications* 9 (3): 864–79.

Thompson, C. 2001. When elephants stand for competing models of nature. In A. Mol and J. Law (eds), *Complexities: Social studies of knowledge practices (science and cultural theory)*. Durham, NC: Duke University Press.

Ticktin, T., and T. Johns. 2002. Chinanteco management of *Aechmea magdalenae*: Implications for the use of TEK and TRM in management plans. *Economic Botany* 56 (2): 177–91.

Usher, P.J. 2000. Traditional ecological knowledge in environmental assessment and management. *Arctic* 53 (2): 183–93.

Western, D., and M. Wright (eds). 1994. *Natural connections: Perspectives in community-based conservation*. Washington, DC: Island Press.

Electronic Resources: http://biotech.icmbu/utexas.edu/botany/perihist.html.

Chapter 10

Integrating Scientific Information, Stakeholder Interests, and Political Concerns

Lawrence Susskind, Patrick Field, Mieke van der Wansem, and Jennifer Peyser

INTRODUCTION

Central to resource and environmental management is the question of what qualifies as fact. Regulatory agencies have responsibility for deciding what new studies are needed, who should conduct them, and what existing data are acceptable sources of information for rule making, permitting, management plans, and other decisions. However, public and private stakeholders have the right to question the credibility of the scientists and the science used, to generate studies of their own, and to challenge the decisions that use this information. These challenges may come in the form of public outcry, public relations campaigns, or even litigation. While such efforts provide an outlet for airing concerns, they tend to add to the overall level of public uncertainty. Often, there is no forum to evaluate what, whether, and how scientific and technical information should be incorporated into specific policy decisions.

No matter the scale, scope, or substance, all environmental and resource management challenges are characterized by the need to cope with scientific uncertainty and make difficult choices about how to value scarce resources. Collaborative decision-making processes seek to involve contending stakeholders in making these decisions together. Their goal is to respond to all, rather than choose among, competing interests and competing approaches to handling scientific uncertainty. Joint fact finding (JFF), as part of a larger consensus-building process, offers a mechanism to bring together decision makers, stakeholders, and scientific and technical experts to build a common base of knowledge to inform resource management decisions. JFF promotes integration across disciplines, across sectors, and across agencies, and allows for the consideration and incorporation of social, cultural, economic, and ecological principles in the formation of environmental and resource management and policy decisions. Thus, integrated resource management is an effort not merely in linking scientific and technical disciplines, but in integrating diverse stakeholder groups, the public included, as well as policy-makers and scientists, into the ongoing process of planning, decision making, monitoring, and evaluation.

GENERAL PROBLEM

Resource and environmental management involves multiple parties with multiple interests, different levels of technical expertise, varying scientific backgrounds, and competing political priorities. Because of these factors, decision makers tend to separate science from other important resource management considerations. Often, they see questions of science as distinct from questions of values. This can blind them to the non-objective decisions inherent in the generation and interpretation of scientific data.

Scientists as well as decision makers prefer to separate science and policy. In the scientific community, 'good science' is often equated with non-political science, so scientists distance their work from the decision-making process. In their efforts to remain objective, they limit their interaction with those in the policy arena before, during, and after their scientific inquiry. Further, it is taboo for scientists to attempt to interpret how their research conclusions might apply to the policy decision at hand. Thus, their involvement ends the moment they hand their report to the agency representative who commissioned it. Scientists who work too much with decision makers or stakeholders risk losing neutrality in the eyes of their colleagues. Unfortunately, by distancing themselves from policy-relevant questions, scientists may also lose credibility with stakeholders.

Rather than working with scientific or technical experts and asking questions about their research or conclusions, stakeholder groups often commission their own studies. This additional research can bring to light new and important questions that help inform policy. However, some parties use science as a means of influencing a policy outcome. By hiding their interests behind a scientific report, some stakeholders are able to push for a policy outcome most favourable to them. This is called *adversary science*. The result is an ever-growing list of 'facts' and corresponding refutations and an equally long list of values that are left out of the conversation. In this way, the separation of science and policy decisions can lead to the manipulation of both.

Consider the following situation. A federal agency is responsible for creating a management plan for a checkerboard of public and private land in the western United States. Parts of this land are being used for ranching and recreation. Other areas include forest land for which there is a permit application for commercial harvesting. The land also serves as habitat for threatened plant and bird species.

Several different private and non-profit groups, in addition to many local stakeholders, have become involved in the management plan decision-making process. There is disagreement over several facets of the plan, including proposals for a new road through the forest, stricter regulation of recreational activities, and habitat conservation measures. There is also disagreement on regional priorities. Some believe that the region should focus on boosting the economy through timber, ranching, or even oil exploration, while others think that the natural environment should be preserved and used as a resource to attract recreationalists and tourists to the region. Some are more concerned with preserving a certain way of life, such as the ranching tradition that has existed for generations in the region. Others are calling for a change in lifestyle that would restrict extractive practices and place a higher value on the region's diminishing environmental resources.

There is no research report that seems to balance all of the scientific, economic, and political concerns expressed by the many different stakeholders. The agency asked contractors and agency scientists to research the region and the possible impacts of different management proposals. They have just received the final reports, which have interesting scientific conclusions but do not really lead to an obvious policy recommendation. Further, some of the stakeholders don't trust the agency scientists or the contractors who were hired to conduct some of the research. Thus, several stakeholder groups have brought forward additional experts to address issues such as timber harvesting methods, species habitat, ranching, and potential effects of the different proposals on the local economy. Each of the experts claims to be neutral and objective in presenting 'the scientific evidence' for the impact of the land management proposals. These experts have never met with each other or with the stakeholders for a systematic review and discussion of the evidence. Instead, they appear separately to defend their work and criticize the assumptions, methods, and findings of other experts. To further complicate matters, experts representing different disciplines, such as biology, forestry, geology, and economics, do not communicate with stakeholders or help them understand how their very different types of data can help inform the management plan.

Stakeholders who are not also technical experts quickly become frustrated and decide that there is no right answer to the question. They may also be frustrated if they have observational data to contribute to the discussion, particularly if it conflicts with technical data. For example, a long-time local resident or a timber worker may be able to comment on the presence or absence of an endangered bird species in a certain area of the forest. However, because such stakeholders are not trained scientists, both technical and non-technical experts in the group may dismiss their contextual knowledge as being neither scientific data nor couched in technical language; some may believe that the stakeholders have ulterior or biased motives (e.g., 'They are saying that to get a predetermined outcome' or 'Those loggers don't care about wildlife').

Perceptions of unequal distribution of scientific resources can undermine the collaborative spirit and lead to a breakdown of the process, or worse yet, to litigation. The likely outcome of this scenario would be a program based on a political compromise within the 'range' of arguments presented by the duelling experts. Additionally, the conversation on regional economic and social priorities would have been left behind in the midst of the focus on scientific conflict.

THE CASE FOR JOINT FACT FINDING

The integration of scientific and technical information into decision-making processes remains a challenge for those involved in resource and environmental management. Even less explored is the integration of the local (or contextual) knowledge, or observational/anecdotal information, possessed by residents or users of natural resources. Consensus-building processes have the potential to resolve not only the value-based conflicts over natural resources, but also the questions of fact that often delay or even overturn environmental decisions.

Consensus building is the process of brokering or facilitating agreement among the representative group of stakeholders involved in any issue or conflict (see Susskind 2000). Consensus building usually includes information gathering (i.e., joint fact finding) and a negotiation process that follows procedures or protocols that the parties themselves help to specify. The outcome usually takes the form of a written agreement that, in an environmental dispute, needs to be entirely transparent (i.e., open to public review). Because of the complexity generated by the number of parties involved and the technical nature of many of the issues under discussion, most consensus-building efforts in the environmental field need to be managed by a highly trained 'neutral' or mediator.

A full consensus-building process includes six key steps:

1. A convenor initiates a possible consensus-building process by asking a neutral party to conduct a stakeholder assessment. Through this assessment, the neutral party identifies stakeholders and assesses their interests, capacities, and potential for reaching consensus-based agreements.
2. The convenor and the stakeholders determine whether or not to proceed with a consensus-building process. If they do decide to proceed, they come to agreement on stakeholder representatives, ground rules, an agenda, a timetable, and selection of a facilitator.
3. If needed, the parties initiate a JFF process to resolve technical and factual questions and help the group focus on the development of feasible options.
4. The parties engage in a process of deliberation in which they create value by generating options or packages for mutual gain.
5. The parties distribute the value that was created by forming recommendations or proposals for agreement. They promote consensus agreements where possible and enable near-consensus alternatives when full consensus is not possible.
6. Appropriate parties are charged with the responsibility for follow-through of the agreement reached. These include responsibilities for implementation, monitoring, and providing opportunities for stakeholders to revisit and revise their agreements as necessary during the implementation phase.

At the end of this process, the group can go back to any of the six steps above as appropriate.

Consensus building has been used in negotiated rule making, permitting processes, habitat conservation plans, forestry management, and many other resource decisions. Federal and state statutes have legitimized the use of consensus building, and agencies at all levels of government have used collaborative processes to facilitate their decision making. Still, many of these processes have ignored or taken for granted the questions of scientific and technical information. There remains a great need to acknowledge and address technical questions in a collaborative manner so that information is gathered, analysed, and incorporated into decisions in such a way that it is credible and useful to decision makers, stakeholders, and technical experts.

Joint fact finding, as an explicit step in a broader consensus-building process, provides for involvement by stakeholders, including non-technical ones, who may be affected by final decisions in the ongoing process of generating and analysing the scientific and technical information used to make those decisions (see Figure 10.1). Joint fact finding recognizes that experts and non-experts have important roles to play in specifying the scientific questions that need to be answered, selecting appropriate methods of inquiry, interpreting findings, and deciding how to handle assumptions and uncertainties inherent in these findings. In this way, joint fact finding increases the transparency of the collection and use of scientific and technical information in decision-making processes. It also allows for a more comprehensive understanding of the policy implications of scientific and technical input.

It is critical that all participants in a JFF process understand their roles. Joint fact finding should involve

- decision makers, or representatives from agencies with decision-making responsibility;
- managers who will be responsible for implementing the decision;
- representatives of stakeholder groups who will be affected by the decision;
- a convenor who can bring these key parties to the table;
- a neutral facilitator who can create a climate conducive to the joint investigation of issues, productive dialogue, and relationship building among participants; and
- scientific and technical experts who interact with and serve as a resource to other stakeholders throughout the consensus process.

Decision makers are those responsible for writing the environmental plan or policy. While the actual decision makers could be defined as the legislators, agency officials, or others who vote or decide on the approval of new regulations, these final decision makers are generally not involved in negotiations. Representatives of agencies under whose purview the new policy will be promulgated are more likely to participate in the rule-making process and therefore in the joint fact finding. This group should be clear about goals and timelines for the policy process and about where flexibility does and does not exist.

Managers are likely from the same agency as the decision makers, but they are responsible for implementing the regulations 'on the ground' from one of the agency's many field offices. For example, a manager of a new rangeland policy would work directly with local ranchers and other landowners to implement grazing regulations or habitat restoration activities, as well as to monitor activities on public lands. Managers are needed in JFF processes to inform other participants about the practicality of potential new regulations under consideration by the group.

Stakeholders are individuals or groups who will be potentially affected by a new plan or policy. A convenor and/or neutral should conduct a conflict assessment to determine the range of stakeholders and which among them should be offered a seat at the table. In environmental management issues, stakeholders often include

THE CONSENSUS BUILDING PROCESS AND THE ROLE OF JOINT FACT FINDING

Convener Initiates a
Consensus Building Process
(A Neutral Prepares a
Conflict Assessment)

↓

Convener and Stakeholders
Decide Whether or Not to Proceed
(If so, Generate Agreement on Stakeholder Reps, Ground
Rules, Agenda, Timetable and Selection of a Facilitator)

↓

Parties Initiate a Joint Fact Finding
Process to Handle Complex Scientific
and Technical Questions

↓

Parties Create Value by Generating
Options or Packages for Mutual Gain

↓

Parties Distribute Value in the
Form of An Agreement
(i.e., Recommendations or Proposals)

↓

Appropriate Parties are
Charged with Responsibility for
Follow Through
(Implementation,
Monitoring and Establish
Dispute Handling Procedures)

↓

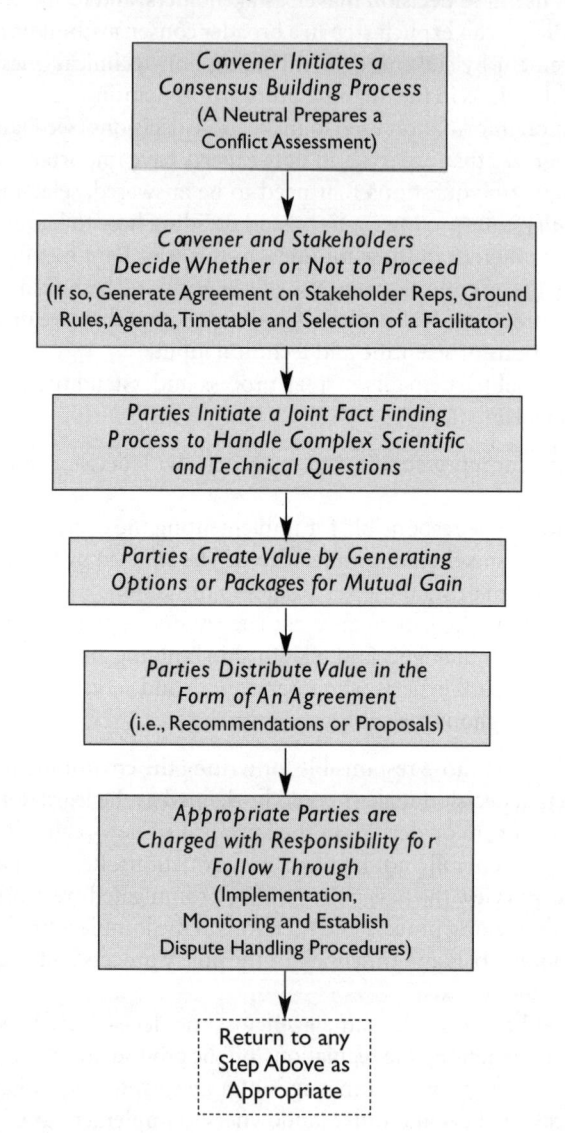

Return to any
Step Above as
Appropriate

FIGURE 10.1 The consensus-building process and the role of joint fact finding

SOURCE: Consensus Building Institute, 2002

advocates hoping to represent the resource in question, such as a river, ecosystem, or a particular plant or animal species. Local community stakeholders may have social or economic concerns, such as the effect of a decision on growth, open space, or the job market. Stakeholders can also include business and industry representatives whose development or delivery of products and services may be affected by the decision. Participants from this group often communicate how regulations will impact resources of their sector, including in terms of time and money.

Joint fact finding also brings in scientists and analysts to work with decision makers, managers, and stakeholders to frame the research question(s), gather data, interpret the data, and communicate the research results. Unlike collaborative efforts where scientists and other analysts are merely consulted, JFF requires these experts to play a central and ongoing role throughout the entire decision-making process. Members of the technical team must be seen as credible to all parties and be able to work in a collaborative situation with decision makers and stakeholders. This does not mean that they should sacrifice objectivity. Rather, they must conduct their work in a transparent fashion, be willing to address stakeholders' questions, and help the group consider how scientific and technical information relates to the policy decision. Further, they must be able to communicate across disciplines with experts involved in other research related to the policy.

In order to bring together decision makers, managers, stakeholders, and experts, a convening group or individual is often needed to provide neutral ground, funding, and logistical support. An agency with policy-making responsibility may serve as the convenor if that is acceptable to the rest of the JFF participants. Generally, convenors take responsibility for identifying and contracting with facilitators. It is advisable for the convenor to confirm that the facilitator is acceptable to all participating stakeholder groups and to set up a procedure for participants to raise and address concerns about the facilitator's performance and/or impartiality.

A qualified facilitator is one of the more important resources in consensus building and joint fact finding. The most effective facilitators are able to work both on the specific tasks of the group and on group dynamics—building a sense of shared purpose, positive working relationships, and camaraderie. Facilitators can do more than just run a good meeting, however, if engaged early in the consensus-building process. If the facilitator enters the process either as the assessor or as the facilitator of initial stakeholder meetings to discuss the results of the assessment and plan the process, then he or she can provide substantial help in process design and process management. In a JFF process, the neutral can help the group identify information sources and experts and facilitate the process of reaching agreement on questions, methods, and interpretation of data. Further, facilitators can promote the legitimacy and effectiveness of the process by

- encouraging effective representation and participation of key stakeholders by helping all participants update their organizations/constituencies regularly;
- helping the group meet its agreed goals as efficiently as possible through careful management of the work plan and agendas for individual meetings;

- identifying resource needs (e.g., funding for consultants, training on technical aspects of forest management or certification) and helping the group meet those needs;
- helping individual participants and the group as a whole with essential steps in the negotiation process, including consideration of each participant's core interests, the creation of options and proposals on specific issues, the development of package agreements, and the crafting of final decisions in light of agreed goals, principles, and criteria; and
- identifying and helping to resolve conflicts among participants, acting as an impartial mediator and problem-solver.

The JFF process has six key steps (see Figure 10.2). In the first step, the convenor, in consultation with a neutral, assesses the need for a JFF process. This assessment should include an identification of what scientific, financial, and human resources will be needed for a successful collaborative inquiry. Convenors should identify some of the data gaps or scientific controversies that could be addressed by joint research. With the help of a neutral in talking to interested parties, convenors should identify a balanced group of stakeholders, decision makers, and managers to work with experts to fill in these gaps. They should also estimate the time and funding needed to convene a JFF process and consider how to cover these costs.

If all parties weigh the costs and benefits of joint fact finding and decide to proceed, convenors can invite the group to the table to begin the process. At this stage, the neutral should assist participants in developing ground rules for working together. Participants should also draft a work plan, which should include discussions on outstanding scientific questions required to inform the policy decision and which experts to involve in researching these questions. Before going further, all parties must understand the different sources of conflict, the questions to be dealt with through joint fact finding, and the other issues that must be considered in the overall consensus-building process. They should also be clear about realistic goals and limitations of their work as it relates to the policy-making process.

Once the JFF group, including the scientific experts, has been assembled, participants must scope the study. Experts can aid other participants in translating their concerns about knowledge gaps or conflicting information into researchable questions. They can also help identify appropriate methods of information gathering and analysis, as well as the costs and benefits of different research methods.

Throughout scoping and the other phases, participants must continue to tie the scientific inquiry back into the policy questions to ensure that their work will be relevant to the decision-making process that will follow joint fact finding. It is also critical for experts to promote transparency in their work by explaining inherent scientific uncertainties and other research limitations.

As the experts conduct the study as scoped by the JFF participants, they should draw on the latter's expertise and knowledge. This could include inputting a manager's observational data into their models or learning about a research site from a local stakeholder. Experts should educate participants about the complexities of their

PREPARE for JFF STEP 1	SCOPE the JFF process STEP 2	DEFINE the most appropriate methods of analysis STEP 3	CONDUCT THE STUDY STEP 4	EVALUATE the results of JFF STEP 5	COMMUNICATE the results of JFF process STEP 6
Understand how JFF fits into consensus building process	Work with stakeholders to draft roles and responsibilities	Translate general questions into researchable question	Undertake the work checking back with constituents	Use sensitivity analysis to examine the overall significance of scientific assumptions and findings	Jointly present findings to stakeholders
Document the interests of all relevant stakeholders	Generate technical questions	Identify relevant methods of information gathering/analysis and highlight the benefits and disadvantages of each		Compare findings to the published literature	Scientists communicate JFF results to various constituencies and policy-makers
Work with a professional 'neutral'	Identify existing information and knowledge gaps	Determine costs and benefits of additional information gathering	Draw on expertise and knowledge of stakeholders	Translate findings into possible policy responses	
	Advise on methods for dealing with conflicting data and interpretations of facts and forecasts	Determine whether proposed studies will enable stakeholders to meet their interests	Review drafts of the final JFF reports	Clarify remaining uncertainties and appropriate contingent responses	Determine if further JFF is necessary
Convene a JFF process				Determine whether and how JFF results have (or have not) answered key questions	

FIGURE 10.2 Joint fact finding: Key steps in the process SOURCE: 2003 Consensus Building Institue, http://www.cbuilding.org

research and check back with them regularly with progress reports, data, and draft findings.

In examining the significance of assumptions and findings and evaluating the study outcomes, participants should use sensitivity analysis. Does changing the parameters of certain assumptions lead to widely different results or to similar results? How much do inputs have to vary to affect outputs? In developing draft and final conclusions, participants should maintain transparency by noting the assumptions, uncertainties, and limitations of the scientific inquiry. They should also compare the results of their studies with the published literature and, if necessary, submit their results for peer review. JFF participants must determine whether all key questions have been answered and how their findings can inform the policy decision.

Communicating the results of joint fact finding is an important final step, as the eventual policy outcome will affect a much larger population than the stakeholders who have participated directly in the process. Participants should prepare key messages from their research findings to share with other stakeholder constituencies and policy-makers. It is important that their communication convey the sense that the research was scoped, conducted, and evaluated in a collaborative manner, and that all members of the JFF team are behind the results. In the communication stage, participants should listen to feedback from other stakeholders and determine whether additional research is needed. If participants conclude that their JFF efforts have yielded the necessary scientific and technical information, they can feed this information into the larger policy-making process.

If stakeholders are able to accomplish these steps jointly, they can dramatically reduce the time and effort spent on debating scientific issues, they can build a shared understanding of the range of uncertainty where there are not definitive factual answers, and they can create a firm scientific/technical foundation for the standard that they recommend. The process can improve collaboration among all stakeholders, facilitating trust among them and reducing the likelihood of 'duelling experts'. Distinct from 'blue ribbon panels' or adversary science, joint fact finding is an inclusive stakeholder process that seeks to generate better and broader understanding of technical issues as well as increased legitimacy for the decision-making process (McCreary, et al. 2001). JFF efforts may also result in a coalition, which could be a major asset in building consensus around non-scientific controversies throughout the policy-making process.

Joint fact finding is particularly suited to integrated resource management and has been used in many environmental decision-making processes, including coastal zone management, watershed management, and facility siting (e.g., McCreary et al. 2001; Ozawa and Susskind 1985; and Jacobs et al. 2003). It is especially useful in settling environmental conflicts when there is incomplete data, when parties make public claims that data are inaccurate, and when stakeholders initially lack the technical capacity needed to participate in decision making (Ehrmann and Stinson 1999).

As much as experts strive for objectivity in their work, practitioners should recognize that the experts' advice and judgments are not always based solely on their technical training and that all technical judgments have a range of value judgments embedded in them (Susskind and Dunlap 1981). Joint fact finding also has mechanisms to deal with non-objective decisions inherent in scientific inquiry. According to Susskind and Dunlap (1981), the following decision points have the potential to be influenced by non-objective criteria:

- choice of professional team members;
- organization of the work plan;
- approaches to coping with uncertainty;
- attitudes towards mitigation;
- approaches to public participation; and
- use of data for and the style of forecasting.

The public should have a role in making these non-objective decisions to help ensure that the research process is consistent with the needs and expectations of the larger decision-making process. Environmental decisions must incorporate the best available scientific information while recognizing the values inherent in such decisions. Consensus-based approaches to decision making provide an opportunity to clarify the scientific and political values that often influence decisions but remain hidden from public view (Ozawa 1991). Joint fact finding can assist in managing the balance between science and values, helping to ensure that scientific information is used in value-laden resource management decisions while leading to effective and stable public policy (Karl and Susskind forthcoming).

Rather than hiding values behind scientific reports, joint fact finding promotes transparency in all phases of research. In this way, the relationship between science and policy can be clarified, and other concerns, such as socio-economic considerations, can be addressed.

Joint fact finding can also complement and strengthen adaptive and integrative management efforts (Ashcraft 2003). *Adaptive management* is the process of making management decisions in light of scientific uncertainty and the dynamic nature of environmental processes. It involves making decisions with the best available information and creating contingencies to be triggered if certain thresholds or unanticipated consequences occur. This allows resource managers to revise and improve plans as they are implemented. Based on feedback from resource managers and data collected from environmental monitoring, adaptive management allows stakeholders to address unintended consequences and grants them the flexibility to improve plans based on information collected during implementation.

As part of a consensus-building process, joint fact finding would make adaptive management a key component of resource decisions by giving stakeholders the opportunity to discuss the uncertainty of scientific and technical information. In dealing with the many unknowns inherent to environmental management in a transparent fashion, stakeholders are able to negotiate how this uncertainty should be dealt with and reflected in the implementation of a resource management plan. While building consensus on the resource plan, stakeholders can also craft monitoring agreements and the criteria for triggering a review of implementation measures. For example, while stakeholders involved in a JFF process to create a rangeland management plan might agree on a certain number of livestock per acre, this number could be increased or decreased based on performance criteria such as surface water contamination and riverbank erosion in grazing areas. The group would agree on a monitoring plan, which could include who would conduct the monitoring of these criteria and how often. They would also negotiate acceptable levels for each criterion, above or below which the group could reconvene to revise their plan.

HOW JOINT FACT FINDING ACTUALLY WORKS, INCLUDING KEY CHALLENGES AND LESSONS

The JFF process varies a little from case to case, depending on the types of issues and the stakeholders involved. But there are a number of typical challenges that stakeholders face when trying to implement this process. We will illustrate below some of these challenges (as well as key lessons) by reviewing the JFF process of the Cancer Incidence Study, which was conducted by the Northern Oxford County Coalition (NOCC), a multi-stakeholder group from Rumford, Maine (the case was originally described in McKearnan and Field 2000).

The Cancer Incidence Study

In February 1991 a popular New England television news show ran a segment called 'Cancer Valley'. The story depicted a rural American community's worst nightmare. It suggested that the people living in Northern Oxford County, Maine, were experiencing extraordinarily high rates of cancer, and it implied that air emissions from a local paper mill might be responsible. Images of local residents walking through cemeteries, with the mill's billowing smokestacks looming in the distance, struck an alarmist note.

The television show clarified (some say exacerbated) a dispute that had been brewing for some time in four rural towns surrounding, and economically dependent on, the mill: Rumford, Mexico, Peru, and Dixfield (hereafter referred to as the four-town area). For many residents of the towns, the television show lent credence to suspicions that a high percentage of residents had contracted cancer. Others opposed the claim that there was a health problem in their community. These residents warned that the label 'Cancer Valley' could tarnish the community's reputation and hinder economic development. Still others were equally concerned that the controversy over cancer rates would force the mill to close, breaking the valley's economic backbone.

Though many residents had very real fears about high cancer rates, little data was available to substantiate or alleviate those fears. Likewise, data on the quality of the air in the four towns were limited to parameters monitored by the mill, hardly a credible source given the controversy. In the absence of trusted information, the controversy seemed sure to produce an extended series of attacks and counterattacks. The regulatory agencies pointed out that the mill was meeting all permitting requirements, including all state and federal regulations for emissions of criteria pollutants. The mill management stated that it had recently invested over $50 million in technologies designed to reduce both the odour and the toxicity of the wastes emitted through the mill's stacks.

The community felt it was stuck, not knowing how to deal with these multiple, emotional, and complex issues. A JFF process transformed the dialogue and allowed the community to begin to resolve the long-standing conflict.

Challenge #1: Scoping the Work
With the help of the US Environmental Protection Agency (EPA) and the state Department of Environmental Protection (DEP), the community convened a stakeholder group to address the controversy. They called themselves the Northern Oxford County Coalition. The 25-member group included concerned residents of the towns (including some employed by the mill), health-care providers, small businesses, the mill management, local and state elected officials (including members of the Maine legislature), and state and federal agencies responsible for protecting human health and the environment.

Once the NOCC was convened, following a stakeholder assessment process, one of the things the members had to decide on in their first meeting was how to proceed. To get a work plan going, the NOCC's facilitators (who knew no more than any NOCC member about how epidemiologists might assess whether there is a cancer problem in a community or how air-quality experts would judge the quality of the air people breathe) took two steps:

1. They went back to the results of the stakeholder interviews completed during the assessment process to clarify what kinds of questions the NOCC might want to ask as part of a JFF process.
2. They got on the phone with expert epidemiologists, toxicologists, and air-monitoring experts to ask how one would go about answering the question of whether cancer rates were unusually high in their community. Specifically, NOCC members wanted to know where to find the necessary data and how long it would take to analyse the information.

Based on this effort, the facilitators presented the NOCC with a set of options on how to proceed:

- they could undertake a study of cancer incidence or respiratory illness;
- they could analyse air quality on the basis of existing data; or
- they could design new monitoring/modelling or lifestyle choices, involving issues such as smoking and diet.

One of the hard choices was what to do first. During the interviews conducted by the facilitators, many stakeholders, even those who disagreed adamantly with one another on other issues, had listed their worry about cancer as their highest concern. However, representatives of EPA and the Maine DEP argued that studying cancer was too difficult because of its multiple forms and the vast uncertainty about its causes. They also pointed out that because cancer has such a long latency period, current cancer rates would not be a good indicator of the valley's present environmental conditions. They suggested that an investigation into respiratory illness might better indicate the current status of the valley's public health and environmental quality. Still other NOCC members, the health-care providers among them, thought that the NOCC should focus first on helping residents see the need for healthier lifestyles. In the end, the NOCC decided to launch its work with perhaps the most controversial and difficult issue at hand—cancer in the valley.

The members agreed to embark on a joint study of cancer incidence. The Maine Bureau of Health had a large, multi-year database of cancer cases in the state, so completion of the study seemed feasible. It would be affordable and doable, but more importantly, it would speak to the strong concerns of the citizen representatives. It seemed that pursuing any other issue would be seen as a diversion if not an outright denial of the problem. Even though there were compelling reasons why a cancer incidence study might not be the most pragmatic choice, the stakeholders would only commit to the process if they felt it addressed their fundamental concerns. Once the coalition had grappled with the mystery of local cancer rates, it might then take on other issues, such as air-quality monitoring, mill emissions, emissions from transportation sources, and/or smoking habits.

Key Lesson: Joint fact finding must be based on the interests of the stakeholders, not merely on what the experts think is the most technically feasible and prudent subject.

Challenge #2: Selecting an Expert

It was clear from the beginning that the cancer study was going to be a complicated and time-consuming task. The first step was to assign the bulk of the work to a subset of NOCC members who cared about cancer incidence, whose background would prepare them for the work, and who were committed to putting in the time needed. A technical subcommittee (TSC) was formed to initiate and oversee the study. As a group, the TSC members represented the key stakeholder groups in the NOCC, including citizens, the union, the mill, and the state and federal public agencies. While there were individuals with scientific and medical backgrounds on the TSC, its members also included local citizens without such background.

The TSC members began by asking two questions: what kind of expert assistance do we need to carry out this study and where can we find it? They quickly realized that one or several people trained in epidemiology would be the best equipped to help. The mill and the union both had their own experts to recommend. The Department of Public Health (DPH) offered to help, and while at first the TSC thought that this was a good idea, it then realized that because of the agency's missteps years ago, its legitimacy had been left wanting in the eyes of the citizens. Thus, the TSC decided that the best route would be to identify a single outside expert chosen by the whole group.

The TSC suggested a few people to contact in the state of Maine. The facilitators volunteered to contact these individuals and to make some additional inquiries. Telephone calls to universities in New England and New York turned up a number of scientists willing to advise the NOCC. Then the TSC convened to review the resumés.

By the end of this meeting, the members decided to recommend that the NOCC retain Dr Bill Barnes (not his real name), a highly qualified epidemiologist from a respected New England university. They reached the decision quickly and efficiently. They also agreed to invite Dr Barnes to the next TSC meeting to present methods for analysing Rumford's cancer rates, as well as to a meeting of the full coalition just one week later. But the plan began to unravel.

As it turned out, and unbeknownst to the mill's NOCC representative, Barnes had given a video deposition as an expert witness in a suit brought against the mill on behalf of sick workers. Just one night before the next TSC meeting, several representatives of the mill called the facilitators and insisted that Dr Barnes be dismissed. They appealed to the facilitators, asking how they could trust this expert. They apologized for not realizing this sooner, but said they had no doubt that Dr Barnes was an advocate for the union and would therefore not provide the kind of non-partisan advice that would make the cancer study credible.

Since this appeal was made at a very early stage of the coalition's work, members were quick to be suspicious of one another's intentions. When the mill's representative brought his request to the TSC meeting, some TSC members were furious, demanding to know why the mill was reneging on an earlier agreement. With Dr Barnes waiting outside, the NOCC members argued about what to do. Some insisted that Dr Barnes was perfectly capable of being neutral and that the mill was stonewalling the initiation of a study to uncover facts about cancer. Others supported the mill's request, noting that it was understandable why the mill was concerned about Dr Barnes's neutrality.

In the end, the subcommittee decided to follow through on the decision to have Dr Barnes give a general presentation on epidemiology at the next NOCC meeting, but they agreed that a new technical adviser should be found immediately afterwards. The facilitators encouraged the committee to talk about criteria that could be used to ensure that members' preferences would be accounted for in the selection process. The TSC agreed with the mill's proposal that the NOCC'S technical advisers should not have had any past involvement in litigation involving any of the constituencies represented in the group. The TSC then requested that each potential adviser be asked to fill out a detailed disclosure form identifying any prior connection he or she might have had with the NOCC's members and affiliated organizations.

Over the next week, the facilitators scrambled to locate advisers who would have the same depth of experience as Dr Barnes but who would be seen as credible by all of the NOCC's constituencies. While discussing the new resumés, NOCC members again raised sharp objections to the mill's eleventh-hour protest. But the group was able to reach consensus on the selection of Dr Daniel Wartenberg, an epidemiologist from Rutgers University. Dr Wartenberg had no prior involvement with any of the parties and was viewed by everyone as having the ability to offer non-partisan advice. He was also generous in his offer to assist the NOCC, expressing interest in participating in a community-based process and only asking for a small stipend and travel expenses.

Key Lesson: The legitimacy of the process is strongly enhanced by a stakeholder group jointly identifying, reviewing, and selecting an independent expert(s) to aid them in their work.

Challenge #3: Refining the Study Design

Even though a cancer incidence study is relatively straightforward, the TSC had to wrestle with numerous assumptions and choices in study design. It began by seeking to refine the purpose for the investigation. Some members wanted to explore the linkages between cancer cases among workers and exposures to toxic chemicals emitted from the mill. Others thought they should first examine the rates of cancer to determine if there were elevated rates worth worrying about.

Given these possible researchable questions, Dr Wartenberg reviewed with the TSC an inventory of potential research methodologies, from a community health survey, to an analysis of local prescription use, to a case-control study. For each, he explained the advantages and disadvantages, including the availability of data, the degree of uncertainty in interpretation, the cost and time involved, and the methodology's ability to answer questions important to the TSC members.

Ultimately, the TSC members agreed that an investigation of cancer incidence using Maine Cancer Registry data would be the best option. This was not an easy decision, as some stakeholders found it difficult to accept that, at least at this stage, the investigation would not pinpoint causes of cancer, but only the rates of cancer as compared to other places.

Further discussion about the merits and limitations of the TSC's approach brought to the surface some of the members' serious concerns about the quality of the data in the Maine Cancer Registry. Some members were worried that cancer cases might have been under-reported in the registry in its early years. With Dr Wartenberg's help, the TSC built into the study design a quick test to help assess the possibility of significant under-reporting. NOCC members also worried that other cases remained undocumented because residents of the four towns had been treated in other states. The TSC's representative from the Bureau of Health played a key role at this juncture, helping the group understand the strengths and limitations of the registry's cancer database.

The TSC also had to determine what the rates in the four-town area would be compared to. After all, rates would only be 'high' or 'low' if they were compared to rates from somewhere else. The group reviewed various options, including comparing local cancer rates to rates in other similar Maine towns (perhaps a mill town), in the remainder of Oxford County, in the state as a whole, and/or in the entire United States.

While some members liked the idea of comparing Rumford cancer rates to those of a similar mill town, the group raised two concerns.

1. If the rates ended up quite similar, did this mean that there was no problem, or did it mean that both towns had elevated cancer rates, perhaps associated with the mills?

2. How would another town feel if NOCC members dredged up that town's data on cancer incidence without their explicit permission?

Dr Wartenberg emphasized the advantages of comparing Rumford rates to a database with a significant number of cases. Such a comparison would increase the likelihood that the results would be statistically significant.

The TSC spent three or four meetings making decisions such as these and eventually developed a methodology for the study, which it brought before the full NOCC membership. The proposed study design entailed an investigation of average cancer rates for 22 different kinds of cancer in men and women for the period 1983 to 1992, and a comparison of those rates to rates in the state of Maine and to a national database, called the US SEER white population database (the Surveillance, Epidemiology, and End Results Program of the National Cancer Institute). The NOCC approved the study design with a few minor changes.

Key Lesson: Give multiple stakeholders the opportunity to wrestle with the hard choices and assumptions one must make in conducting a study that is scientifically credible and salient to the questions and concerns of stakeholders.

Challenge #4: Interpreting the Data and Writing the Report

When the data arrived, Dr Wartenberg prepared statistical tables for the TSC (and eventually the NOCC) to review. The tables showed that the rates for all cancers combined for both males and females in the four-town area were elevated when compared to those for Maine as a whole. It showed that males had a statistically significant elevated rate for cancers of the respiratory system, male genital system (primarily the prostate gland), and lymphatic system (lymphomas). Females were shown to have a statistically significant elevated rate for cancers of the endocrine system (primarily thyroid) and the colon. A number of other types of cancer were elevated in females but not to a statistically significant degree.

While the facts seemed straightforward enough, the next step was to interpret the data and to summarize the results in a report for the full NOCC and finally for all residents of the four towns. This was perhaps the most challenging part of the Cancer Incidence Study. In order to develop the content of a cancer incidence report, the TSC had to agree on how to present the data. But it soon became clear that interpreting the data was not an objective exercise. For example, some NOCC members thought that the report should say that some of the higher-rate ratios (a number that compares the local cancer rate to the state or national rate) warranted concern and further investigation. Other TSC members were equally insistent that epidemiologists would not typically be concerned about ratios unless they indicated cancer rates two to three times the state average. They pointed out that those elevations less than two times higher were just as likely to be 'noise' as they were to be indicators of real problems, especially with such a small data set (thousands of people rather than millions). This led to a long discussion about whether the TSC should include in its report a benchmark that signalled when the community ought to be concerned about a cancer. Some thought that any rate ratio that was elevated should be of concern. Others felt that the subcommittee should not raise concern unless the rate ratio was two to three times greater than expected. And still others argued that the TSC should just present the numbers and let the readers decide.

Reaching an impasse on this issue, the TSC decided to have its draft report peer reviewed. If the subcommittee couldn't resolve these questions, perhaps advice from three professional epidemiologists would help. Interestingly, these experts each had different answers to the TSC's question about when a community should be concerned about a specific rate ratio. In fact, the peer reviewers' responses reflected the range of opinions held by TSC members.

Turning to peer reviewers did not produce agreement. But TSC members discovered that what seemed a local, highly partisan dispute was also a disagreement among scientists across the country. This helped the subcommittee gain an appreciation for the difficulty of drawing precise, universally shared conclusions about technical issues. From this new vantage point, they could agree to the facilitators' recommendation that they abandon the quest for a singular consensus on how to interpret the rate ratios for the community and instead agree to describe the range of views among them in the body of their report.

Key Lesson: In joint fact finding, the meaning of data, analysis, findings, and conclusions is discussed, deliberated upon, argued over, and ultimately determined by the stakeholders themselves. Conclusions and implications are drawn jointly.

Challenge #5: Communicating the Results

During the time the TSC was collecting data, analysing it, and writing the cancer incidence report, the NOCC met to hear reports from the subcommittee about its progress and to provide input. Still, when the final report was presented to the full coalition, it was apparent that additional negotiations were needed.

Some NOCC members were frustrated that the report was only an incidence study and did not include any analysis of what might be causing the elevated rates. Others were frustrated that the report didn't make a clear statement about whether or not there was a cancer problem in the valley. For those who had long been convinced that cancer rates were sky high in the four towns, it was hard to fathom why the TSC couldn't say that there was a confirmed health problem wherever there was a local cancer rate that was statistically significantly higher than the same cancer's rate statewide. At this point, the TSC needed to assist Dr Wartenberg in educating the larger group. TSC members explained that statistical significance is just one of the many factors that scientists weigh when evaluating cancer incidence, and pointed out that scientists often disagree in their conclusions about what to be concerned about. The TSC banded together at that point, seeking to persuade the NOCC that what was important was reaching agreement on what all of the stakeholders could do to follow up on this incidence data, both with further research and with concrete actions to improve public health.

Despite frustrations on the part of a few members, the NOCC ultimately did reach agreement on the report, 'A Report on Cancer Incidence in the Rumford Maine Area.'

The NOCC also decided to leave the letter outlining the concerns of one TSC member attached to the final document. The main motivation for working hard to bridge remaining differences was a shared desire on the part of the NOCC members to get the cancer incidence data out to residents of the four towns. Copies of the final report were distributed informally by NOCC members and placed in the town libraries. The final report was sent to health-care providers, community organizations, and the local hospital, where the NOCC sponsored a 'Grand Rounds' on the results. All attending physicians attended a briefing by Dr Wartenberg and the committee on the results. A summary of the study was also printed as an insert in 7,000 copies of the local newspaper.

Key Lesson: In joint fact finding, the direct participants in the process become the best advocates for the study's methods, data, and conclusions.

Conclusion

The NOCC Cancer Incidence Study did not find a smoking gun that identified the cause of the increased rates in the valley (or failed to identify it, for that matter). However, the incidence study did bring to the community credible, legitimate information on a highly controversial and painful issue. Furthermore, by working through the details and limitations of the study, the members of the community (or at least of the NOCC) discovered that they could work together, that trusted information could be produced, and that information was not the end all and be all. In the end, the NOCC decided to spend its remaining dollars, not on further study (which was quite possible and might have been useful), but on creating a new non-profit organization broadly dedicated to public health. The River Valley Health Communities Coalition was formed when the NOCC disbanded, and it has been in operation since 1997.

Key Lesson: In joint fact finding, participants learn first-hand the power and limitations of data and study.

CHALLENGES OF IMPLEMENTING JOINT FACT FINDING

Besides the challenges illustrated above through the Cancer Incidence Study, there are several other typical challenges that parties may face in a JFF process. Below, we describe these other challenges and some possible responses to them.

1. *There may be resistance from potential convenors (agencies) who fear that their authority will be undermined in a JFF process.*

Many resource managers agree in principle with the importance of public involvement and collaboration. However, it is unclear that there is widespread agreement on what constitutes the right amount of public involvement, particularly when it comes to questions of scientific and technical information. Resource managers (and their technical staff) feel that the wrong questions will be asked, the scoping of the studies will be gamed to ensure a preordained outcome, the public won't understand, the clarity and purity of science will be undermined by the meddling of stakeholders, or the outcome will never be reached owing to conflict and stalemate among diverse interests.

Response: Because of this resistance, a contract should clearly define the appropriate roles and responsibilities of each of the stakeholders, including the resource managers. Clear timelines, available resources, and work plans should be developed jointly to ensure not only transparency but also good project management. Furthermore, showing concrete examples of the results of successful joint fact finding might be helpful. It should also be made clear to all stakeholders that if, after an initial stakeholder assessment, they do not agree with the recommendations for moving forward, they have no obligation to stay involved. And, importantly, all stakeholders, including the convening agency, have the right to withdraw from a JFF process at any point if their interests are not met.

2. *Some stakeholders might be reluctant to join a JFF process because they think that agreeing to participate in joint fact finding will lead to co-optation.*

Some stakeholders may think they are better off staying on the sidelines and trying to block or undermine the official process. They may think that negotiating with groups with whom they have traditionally disagreed will lead to unacceptable compromises. Some stakeholders may also be worried that they do not have the negotiating skills or institutional capacity to participate effectively or to achieve their goals.

Response: Joint fact finding is part of a voluntary process. Stakeholders should understand that a JFF group as a whole will determine the decision rule—consensus involves seeking unanimity but settling instead for overwhelming agreement when every effort has been made to hear and respond to the concerns of all participants. Thus, stakeholders cannot simply be cut out of the process (especially regulatory agencies). Further, participation does not preclude legal or political options. Joint fact finding is intended to expand, not limit, stakeholders' opportunities to have their needs met. Importantly, joint fact finding can help diverse parties pool their technical and financial resources so that more, rather than less, technical work is accomplished.

The institutional capacity to participate is an important issue. Joint fact finding requires that stakeholders commit time to the process, but it recognizes that different parties have different skills. Participation brings many benefits. For example, stakeholders gain inside access to a great deal of information that might otherwise be unavailable to them. Additionally, participants are given the chance to build their technical knowledge and capabilities. Stakeholders can request training in negotiating and consensus building before they begin joint fact finding, skills that can be applied to future resource management issues they may encounter. They can also request to be briefed on any number of technical issues, from hydrology to risk assessment to population ecology. And, if resources permit, stakeholders can request that the convening agency pay for a technical adviser to support their participation.

3. *There may be resistance from key powerful parties, particularly private sector or non-convening agencies, that expect to use back channels to influence outcomes and worry that the transparency of joint fact finding puts them at a political disadvantage.*

Response: For those who find that backroom deals and connections to the powers-that-be are their chief means of doing business, yes, joint fact finding is a problem. Technical studies undertaken by diverse groups are less open to manipulation, suppression, and controversy. But a group that wants others to negotiate in good faith needs to be willing to negotiate in good faith itself. Stakeholders should keep in mind that transparency applies to all parties and that being upfront about needs and goals can save them valuable time, time that is often spent guessing what other parties really want. And, as always, if a group is unsatisfied with the outcome of joint fact finding, it is not required to sign any agreement or give up efforts to have its needs met through other political or legal channels.

4. *Lack of money—what's the point of starting a JFF process if you lack the funds to do the necessary work? Perhaps it is better not to pretend that joint fact finding is possible if some of the necessary resources are not available.*

Response: Collaborative processes such as joint fact finding are sometimes a magnet for additional funds. The gathering together of a diverse group of stakeholders is a powerful signal to convenors and potential funders that there is a willingness to solve the problem at hand. Funders understand that money spent on consensus building will not make matters worse, and they tend to like the idea of a single pooled effort, rather than competing efforts, to conduct research, create models, analyse data, and undertake other technical activities. Whatever money is or is not available, it is still better to have the product generated jointly so that the results will be credible to all parties and used in a manner that actually improves the resource management decision.

5. *Scientists don't know how to participate in JFF. Because they do not want to be viewed as biased, they keep thinking that they should mail in their findings and stay as far away from policy discussions as possible.*

Response: Scientists are beginning to learn about new ways of being involved and making their research relevant to current policy questions. Capable scientists understand that they can help frame questions and issues, and can design methodologies that address questions of pertinent public concern. Many scientists are learning how to work with, and not just for, clients that include not only agencies, but agencies and their stakeholders. Joint fact finding does not require scientists to advocate policy. Rather, it helps them learn how to assist groups working collaboratively to connect the results of JFF with policy choices in reasonable ways. In fact, joint fact finding can give scientists the opportunity to advocate for and undertake work that has a much higher chance of actually being used in policy-making, because the research conducted through joint fact finding builds ownership and gains so many more supporters. And for scientists who complain that policy-makers and the public 'just don't get it,' joint fact finding can be an important means of building knowledge and confidence in technical and scientific work.

CONCLUSION

Joint fact finding, as part of a larger consensus-building process, offers a mechanism to bring together decision makers, stakeholders, and scientific and technical experts to build a common base of knowledge to inform more integrative resource management decisions. Joint fact finding promotes integration across disciplines, across sectors, and across agencies, and allows for the consideration and incorporation of social, cultural, economic, and ecological principles in the formation of environmental, resource management, and policy decisions.

Regardless of the political party in power, everyone calls out for good policy decisions based on sound science. However, in the words of one US Geological Survey scientist, the problem has been that sound science is something 'invoked by all and listened to by none.' In order for science to be sound, it must be credible, legitimate, and salient. Joint fact finding offers stakeholders a real opportunity to meet this three-pronged policy test of sound science. Whether it be in ecosystem management, integrated resource management, epidemiology, or endangered species habitat protection, joint fact finding can be an effective means to harness science for good policy-making.

REFERENCES

Ashcraft, C. 2003. Applying adaptive management principles to the Cape Wind Controversy. MIT-USGS Science Impact Collaborative Working Paper Series. http://web.mit.edu/dusp/epg/music/pdf/ashcraft.pdf.

Ehrmann, J.R., and B.L. Stinson. 1999. Joint fact-finding and the use of technical experts. In L. Susskind, S. McKearnan, and J. Thomas-Larmer (eds), *The consensus building handbook: A comprehensive guide to reaching agreement.* Thousand Oaks, CA: Sage Publications.

Jacobs, K.L., S.N. Luoma, and K.A. Taylor. 2003. CALFED: An experiment in science and decisionmaking. *Environment* 45:1.

Karl, H.A., and L.E. Susskind. Forthcoming. Joint fact finding—Integrating science into value-laden societal decisions. *Environment.*

McCreary, S., J. Gamman, and B. Brooks. 2001. Refining and testing joint fact-finding for environmental dispute resolution: Ten years of success. *Mediation Quarterly* 18 (4).

McCreary, S., J. Gamman, B. Brooks, L. Whitman, R. Bryson, B. Fuller, A. McInerny, and R. Glazer. 2001. Applying a mediated negotiation framework to integrated coastal zone management. *Coastal Management* 29:183–216.

McKearnan, S., and P. Field. 2000. The Northern Oxford County Coalition: Four Maine towns tackle a public health mystery. In L. Susskind, S. McKearnan, and J. Thomas-Larmer (eds), *The consensus building handbook: A comprehensive guide to reaching agreement.* Thousand Oaks, CA: Sage Publications.

Ozawa, C.P. 1991. *Recasting science: Consensual procedures in public policy making.* Boulder, CO: Westview Press.

Ozawa, C.P., and L. Susskind. 1985. Mediating science-intensive policy disputes. *Journal of Policy Analysis and Management* 5 (1): 23–39.

Susskind, L. 2000. A short guide to consensus building. In L. Susskind, S. McKearnan, and J. Thomas-Larmer (eds), *The consensus building handbook: A comprehensive guide to reaching agreement*. Thousand Oaks, CA: Sage Publications.

Susskind, L.E., and Louise Dunlap. 1981. The importance of nonobjective judgments in environmental impact assessments. *Environmental Impact Assessment Review* 2:335–66.

CHAPTER 11

Dis-integrating Equity: Sustained-Yield Forestry and Sustainability in Vallecitos, New Mexico

Carl Wilmsen

Gavino Alire has mixed feelings about the Vallecitos Federal Sustained Yield Unit (VFSYU). The Forest Service's plan seems as though it could benefit the community as long as the agency follows through and doesn't put more restrictions on current community uses of the land. The plan conforms to new congressional legislation stipulating that national forest lands be managed for the benefit of the local communities. Logs from the forest will be processed in Vallecitos, and wood product manufacturing facilities will be established locally to create even more jobs. There's no doubt the community can use the jobs, and the Forest Service will use the latest science to manage the forest to ensure that it remains in good condition. Although skeptical of the Forest Service's true motives, Gavino feels that this project is worth a try (USDA, Forest Service 1947).[1]

Does this scenario sound familiar? Could this be public reaction to the latest proposal to integrate community needs, poverty alleviation, and environmental concerns in a small village in Latin America? Actually, the year in question is 1947 and the US Forest Service was proposing to establish the Vallecitos Federal Sustained Yield Unit in the Carson National Forest near the town of Vallecitos, New Mexico (Figure 11.1). Integrating environmental, economic, and social goals, the unit was intended to implement a management plan that would restore, improve, and conserve local resources, alleviate poverty among residents of nearby towns, and promote the stability of those communities.

These goals bear a striking resemblance to the goals and objectives of many contemporary efforts to achieve sustainability. So what was different about sustainability in this case? The conceptualization of sustained yield that underlay the VFSYU's management plan emphasized integrating economic and environmental management and, to the extent that social and economic equity was considered at all, assumed that community stability was a natural by-product of sound economic management. This assumption has resulted in an inequitable distribution of the costs and benefits of the VFSYU's management plan. Local area residents have borne the cost of the

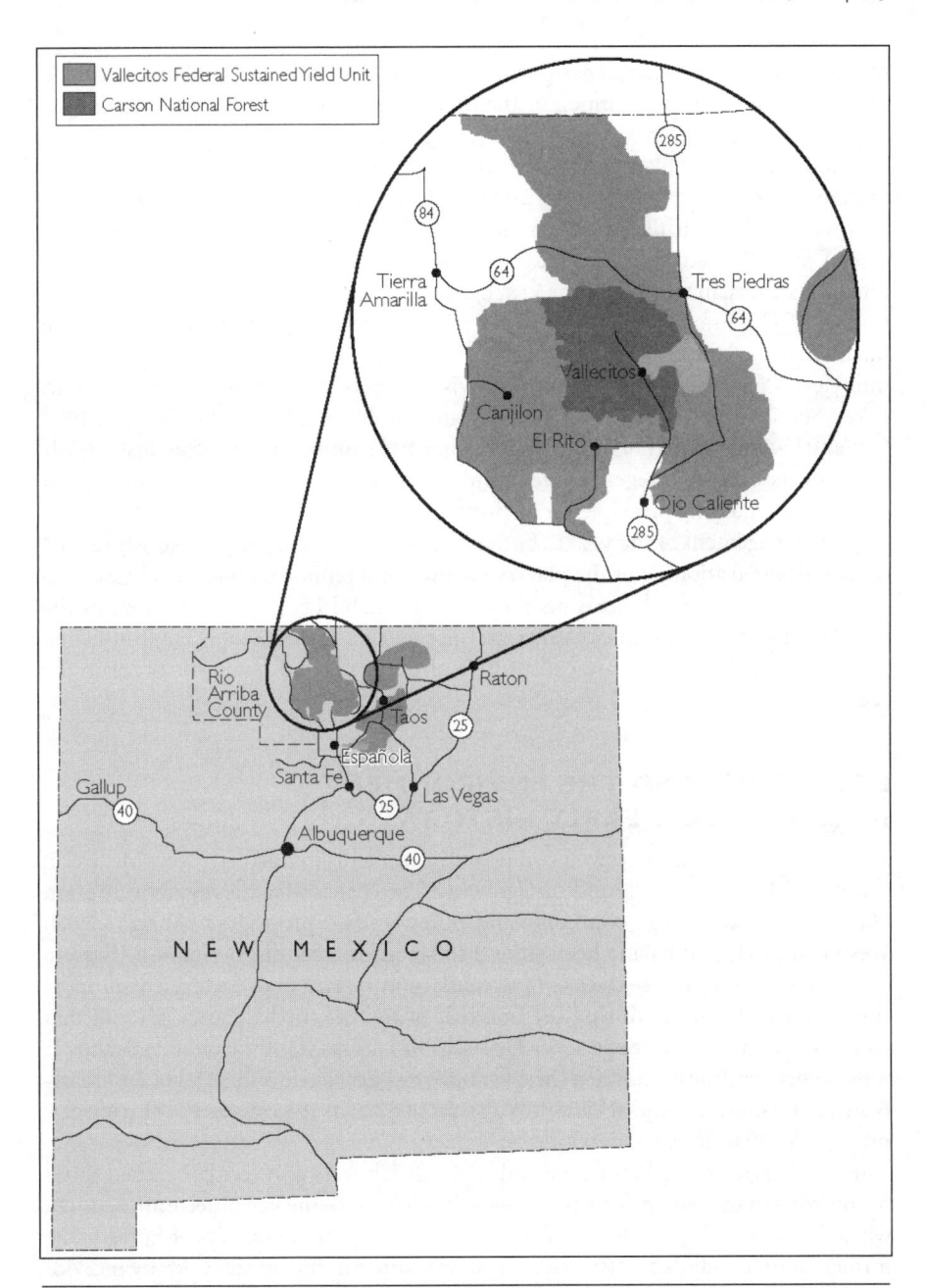

FIGURE 11.1 Location of the Vallecitos Federal Sustained Yield Unit

reduced numbers of livestock permitted to graze on national forest lands, have worked for substandard wages for much of the unit's history, and have fought to hold the Forest Service and forest industry accountable to policy objectives.

The sustainability plan promised to change such outcomes with its focus on integrating economic, environmental, *and* social concerns. This raises the question of how well efforts to achieve sustainability actually do bring equity into the mix. The case of the VFSYU suggests that much remains to be done in this area. Despite good intentions, the management of the VFSYU has reproduced the subordinate social positioning of Hispano residents relative to other interests and has contributed to the maintenance of the region's racial order. Drawing on archival research and on the interviews I conducted with Hispano activists, loggers, and ranchers, as well as with Forest Service officials, environmentalists, and lumber company officials, from 1994 to 2000, I will demonstrate how this tendency has continued even after sustainability, with its call for integrating economics, ecological science, and social equity, emerged as a guiding principle for environmental management.

The management of the VFSYU, both past and present, has simultaneously turned on the social positioning of the players involved and reinforced that social positioning. This is evident in the reasons the unit was established, in the structure of the unit's management, and in contemporary approaches to sustainable environmental management, all of which are constitutive of the racial and economic order in the region.

ECONOMICS AND THE ENVIRONMENT IN THE VFSYU'S EARLY HISTORY

Forest Service officials perceived two pressing problems in northern New Mexico in the mid-1940s: overgrazing and rural poverty. The Hispanos in the region were poor. Having had their village commons and other lands expropriated (Ebright 1994; Westphall 1983) and having been subject to labour market discrimination (Barrera 1979; Deutsch 1987) since the late nineteenth century, Hispanos endured disproportionate rates of poverty during the first half of the twentieth century, a trend that continues to this day (Carlson 1990; Deutsch 1987). This was the case in Vallecitos as well. In his feasibility study for the VFSYU, Forest Service forester David Scott estimated that family income in Vallecitos was about $280 in 1935 (Scott 1947), a low figure even for that time.

Forest Service and other federal and state officials were also very concerned about overgrazing in the region. This had been a problem since the late nineteenth century, when the arrival of the railroad made the commercial production of beef cattle on the former common lands an attractive business venture in the territory. Many individuals and private corporations rushed into the cattle business in the hope of realizing huge profits. Stocking levels well above the carrying capacity of the land resulted in a serious overgrazing problem (deBuys 1985; Rothman 1989; Westphall 1983). The 1930s saw the expression of official concern for the persistence of overgrazing and poverty in a number of government studies, reports (Cooperrider and Hendricks

1937; USDA, Soil Conservation Service 1939), and the Roosevelt administration's New Deal programs.

By the 1940s, the problems had not abated, and one of the areas of concern to Forest Service officials was Vallecitos and the surrounding Carson National Forest lands. Agency officials sought a remedy to the dual problem of entrenched poverty and unrelenting overgrazing in the Vallecitos area in the provisions of the Sustained Yield Forest Management Act (SYFMA) of 1944, which authorized the creation of sustained yield units.

To deal with the overgrazing problem, the agency planned a livestock reduction program. Agency officials worried, however, that implementation of such a program would reduce the already poor local population to penury. In creating the VFSYU, they reasoned, they could integrate economic and ecological approaches to address the overgrazing problem and could offset individual losses in livestock production with wage-earning opportunities in an expanded local lumber industry (USDA, Forest Service 1947; Scott 1947; Clary 1987).

This reasoning was rooted in a new concept of sustained yield upon which the SYFMA was based. Although sustained yield had meant holding harvest levels to the annual incremental growth of the forest, the SYFMA was based on an understanding that shifted the focus of management from maintaining the forest to maintaining the sawmill (Schallau and Alston 1987). David Mason, a chief architect of the act, reasoned that controlling the supply of lumber would yield better prices, thereby making lumbering more profitable. This in turn, he argued, would ease competition among lumber firms and encourage more careful management of forests. Mason further argued that this strategy would enable lumber companies to remain indefinitely in the communities in which they were established, thus putting an end to the boom-and-bust cycles in which companies left communities after harvesting all the economically available timber, leaving behind broken families, problems of substance abuse, and other social problems that typically plague a suddenly unemployed labour force (Mason 1927).

Sustained-yield units embodied this logic—this integration of economic and ecological goals—in their design and management. To encourage the establishment of a permanent industry in a local area, the SYFMA authorized the Forest Service to designate specific companies as principle responsible operators for the local unit. The Act also authorized them to offer these designated operators timber on the unit at appraised value, without competitive bidding, required that they establish and maintain sawmills in the local area for at least primary manufacture of rough-cut lumber, and stipulated that they hire the majority of their labour from the local community (SYFMA 1944). In essence, designated operators were granted monopoly access to federal timber in exchange for their compliance with certain requirements intended to contribute to the economic development of the local area.

Although the SYFMA allowed room for interpretation in its implementation, the law's emphasis on timber, as opposed to other forest resources, as well as the Forest Service officials' assumptions in applying it, had major implications for the Vallecitos community:

1. Forest Service officials have interpreted the law to mean that timber is the resource to be planned for on the unit, and Vallecitos-area residents have responded by pursuing timber-harvesting opportunities.
2. Forest Service officials favoured a large company, as opposed to several smaller ones, in establishing the unit, which meant that the agency had to depend on large, well-capitalized firms to implement its plans.
3. The preference of a large company over a number of smaller ones meant that the only expected contribution of the local population to the unit's operation was as a labour force.
4. Agency officials' notions about Hispanos as an ethnoracial group reproduced the racial order in the region.

That the Forest Service had a large company in mind for implementing its plans for the VFSYU is evident in the closure of one of several small sawmills operating in the vicinity at the time. The Jarita Mesa Lumber Company's sawmill had been in operation for some 20 years near the town of Petaca (Figure 11.2) when in 1949 the Forest Service asked it to move to another locality. To make the unit more attractive to potential designated operators, who would be required to make a substantial investment to meet the unit's remanufacturing requirements, agency officials wanted to add the allowable annual cut available to Jarita Mesa Lumber Company to the annual cut it was planning for the unit. In response to protests from local residents, who had asked US Senator Dennis Chávez to intervene, Regional Forester C. Otto Lindh wrote: '[I]t is our sincere belief that a good plant at Vallecitos with drying yard, planing and remanufacturing will be of far greater benefit to the people of Vallecitos, Canon Plaza and Petaca *than the present small mills* cutting only rough green lumber to be sold and processed elsewhere' (Lindh 1950, emphasis added).

The planing and remanufacturing facilities to which Lindh referred were intended to increase employment opportunities through the establishment of value-added production.[2] While establishing value-added manufacturing facilities is standard economic development practice, the Forest Service's view that a relatively large operator from outside of the area was required to achieve the unit's goals positioned local residents as a labour force, with nothing other than their labour to contribute to the unit's operation.

The notions Forest Service officials had about Hispano life ways and land-use practices reinforced this positioning by delegitimizing Hispano environmental knowledge. People in the agency viewed Hispanos as inept and ignorant land managers and considered their way of life backward. For example, in a 1943 memorandum to Forest Service Associate Chief Earle H. Clapp, signed only with the initials C.L.F., the author states:

> What to do with the underprivileged Spanish-American groups like those in northern New Mexico presents a problem that better heads than mine have failed to crack. These people, like the people in Puerto Rico and in some at least of the Latin-American countries, are like deer in certain respects; i.e., they will

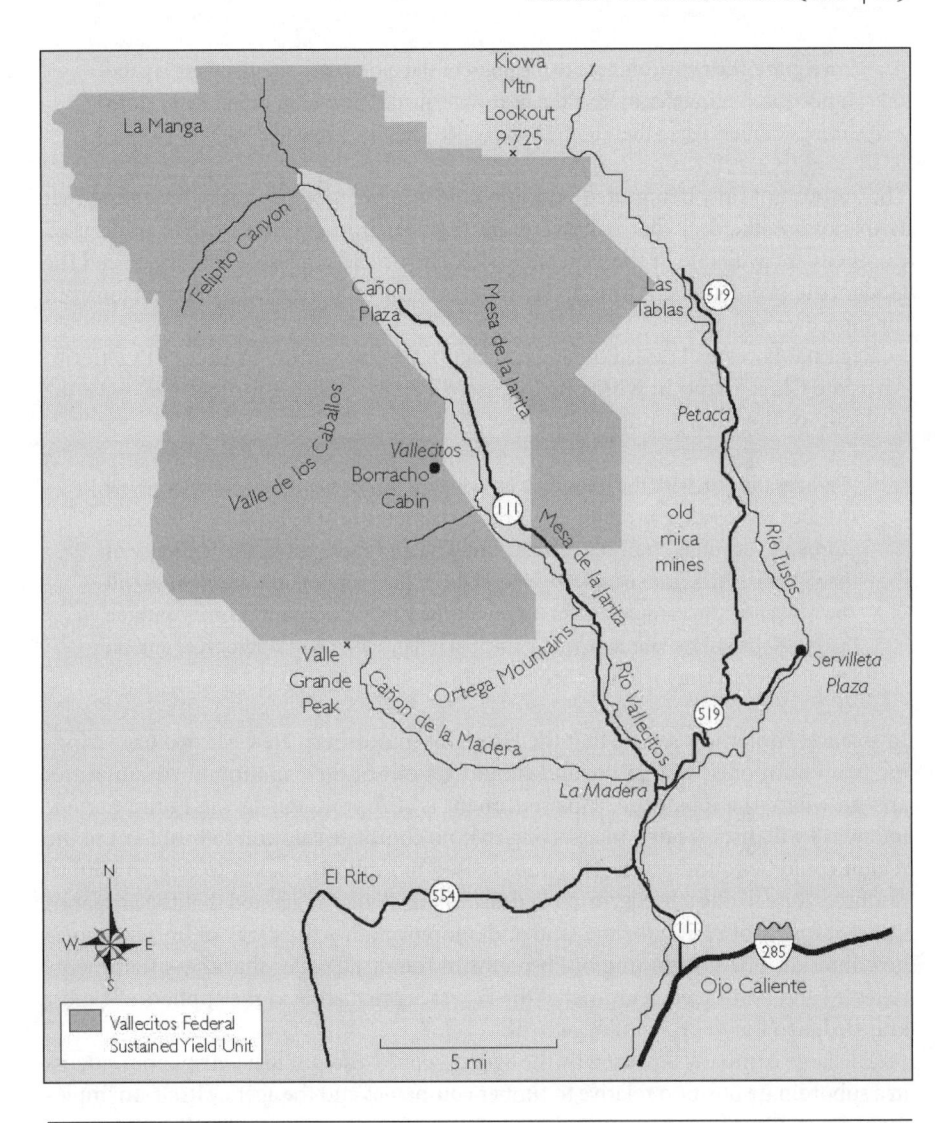

FIGURE 11.2 The VFSYU and Surrounding Communities

'overgraze' their environment to the point of starvation even though there is good under-used resources over in the next watershed. They do not migrate in sufficient numbers when their home locality becomes overcrowded. (C. L. F. 1943)

The author not only denigrated Hispano land-use practices, but also, by stating that Hispanos are like deer that will overgraze their environment even when underused resources are available in the next watershed, he or she suggested that they need the guidance of professional White land managers because, like deer, they cannot look out for their own welfare.

Regional Forester Frank Pooler expressed a similar sentiment in a 1943 letter to Associate Chief Clapp in which he discussed Forest Service efforts to deal with the problems of the rural population in New Mexico.

> Perhaps the war, with the training it is giving the young men from Northern New Mexico will result in breaking down their isolationism. At least I am hopeful that through vocational training, better schooling and more travel the tendency of the Spanish-American people to live and die in their mountain communities will be weakened. No complete solution can be found for this area if people continue to multiply and remain in it with principal reliance on natural resources and on rural or forest employment. (Pooler 1943)

In essence, Pooler was saying that the Hispanos in northern New Mexico were clinging to an outmoded way of life and should get out of their mountain communities and go where the jobs were. This sentiment is still prevalent in the Forest Service today, as I will discuss later. In asserting that no complete solution to problems in the region could be found unless Hispanos ceased to multiply, ceased to rely on natural resources for earning a living, or moved elsewhere, Pooler suggested that the Hispanos were incapable of caring for the land and, moreover, that their way of life was somehow inferior. The positioning of Hispanos in the racial order that these sentiments reflected and reinforced, legitimated the Forest Service's role as the appliers of expert knowledge to the VFSYU's management.

The large company favoured by the agency would also put local Hispano residents in a subordinate position relative to timber companies and the agency itself. To implement its vision of a large mill with remanufacturing facilities, the Forest Service depended upon a large, well-capitalized firm that could afford the initial investment, had access to credit, and had sufficient resources to maintain steady production in the event of equipment failure or other contingencies. Clearly, the poverty-stricken residents of the Vallecitos area did not meet these qualifications, although they did have experience in logging and milling with the several small mills operating in the area, as well as the experience of middle-aged and older men who had worked in the industry in the railroad logging days of the 1920s.

The advantages that the agency's dependence on large, well-capitalized companies imparted to the companies is evident in the many concessions agency officials made to the unit's designated operators from the very beginning. Colorado businessman A.W. Connery and his partners actively sought designation as the unit's official operator for their newly formed Vallecitos Lumber Company when Forest Service officials were first contemplating the creation of the unit. They immediately expressed concern that the allowable annual cut of 400,000 board feet that the Forest Service had set for the unit was too small to justify the investment in a sawmill and in the remanufacturing equipment required by the agency, as stipulated in correspondence with then forest supervisor Louis F. Cottam (Cottam 1946). Connery was evidently successful in pressing his case, for Cottam set the allowable annual cut at 1.5 million board feet when the unit was finally created in 1948. This was indeed a notable change, since the forester who conducted a feasibility study for the unit in 1947 estimated that the annual harvest the unit could supply was smaller than the estimates in the 1941 management plan for the area upon which Cottam had presumably based his original allowable annual cut of 400,000 board feet (Gross 1953; Scott 1947).

This scenario was to repeat itself for every designated operator that operated on the unit until local Hispano-run entities were designated as operators in the 1990s. In 1952 J.L. and J.W. Jackson's Jackson Lumber Company became the designated operator. In negotiating the terms of their contract with the Forest Service, the Jacksons expressed the same concerns about the allowable annual cut being too small as Connery had some six years earlier (Jackson 1952). Again Forest Service officials accommodated the requests of the designated operator; they not only increased the allowable annual cut, but also, in response to the Jacksons' entreaties, increased by 28,000 acres the acreage from which timber could be sold to the Jacksons. While the unit was due for its legally required decennial review at this time, it is more than coincidental that the Forest Service chose this time to scour the forest with the latest timber-inventorying technology (Wall and Egan 1952). The increases in annual harvest and acreage appeared contrary to the unit's original management objectives to forester L.S. Gross in the Forest Service headquarters in Washington, but he nevertheless concluded that the new management plans were sound and approved them (Gross 1953).

Forest Service officials did not stop there, however. Reasoning in 1952 that the annual harvest levels were too low to support such production, Regional Forester Lindh noted that refinement beyond the drying and finishing of lumber was uncertain (Lindh 1952). By this Lindh meant that remanufacture of the milled lumber into new products was unlikely at the Vallecitos mill. His prediction quickly became reality when agency officials eliminated the value-added manufacturing requirements at the Jackson mill.

With these actions, Forest Service officials substantially abridged their original objectives for the unit. In an effort to assure the economic success of the unit's designated operator, they adjusted their appraisal of the unit's timber resources and eliminated a major portion of their program for achieving community stability.

The approach to sustained yield that Forest Service officials adopted thus legitimated the agency's agenda and reproduced the racialized social order in northern New Mexico. It positioned lumber firms in such a way that they could effectively negotiate the terms of forest management with the Forest Service, and it denied Hispano residents the politico-economic position—as well as the discursive legitimacy—necessary to do the same.

The provisions of the SYFMA and the structure of the VFSYU's management did, however, provide opportunities for the local residents to exercise power. Since the law stipulated that the management of sustained-yield units should promote community stability, and since timber sale contracts stipulated that designated operators were to hire 90 per cent or more of their labour from the local area, Vallecitos-area residents organized an informal, community-led organization—the Vallecitos Federal Sustained Yield Unit Association—to hold the Forest Service accountable to the law as well as to its own management directives. Moreover, the local residents and others hired by the Jacksons had the power to withhold their labour. And this they did in a strike they initiated in July 1955.

Ultimately, their efforts had little effect. The Jacksons took a recalcitrant stand, refusing to comply fully with the local labour-hiring requirements or to recognize the labour union that local residents organized and led.[3] The Jacksons fired the strikers and hired scabs, actions that led to violence. The strikers beat up the scabs and, in at least one instance, ambushed and fired upon them with rifles (fortunately no one was seriously injured). Although the union called off the strike in November, disputes among the Jacksons, local residents, and the Forest Service continued to rage until May 1957, when the Jackson sawmill burned to the ground and the Jacksons left Vallecitos for good.

THE VFSYU AND SUSTAINABILITY

With the departure of the Jacksons, the VFSYU was dormant for 15 years, having no designated operator. Then, in 1972, the Duke City Lumber Company was designated as the unit's operator, and before long it reproduced the social relations that had prevailed in Vallecitos during the 1950s. Hispano residents, seen once more only as a labour force, revived the VFSYU Association to again hold the Forest Service accountable. In the mid-1980s, however, local residents involved themselves in the new forest planning processes mandated by the Forest and Rangelands Renewable Resources Planning Act of 1974 and the National Forest Management Act (NFMA) of 1976. While this involvement led to short-term improvements in the situations of a number of Hispano residents, the direction management has taken in the late twentieth and early twenty-first centuries has once again reinforced the subordinate social positioning of Vallecitos-area Hispanos—despite the emergence of the sustainability concept with its call for integrating economics, ecological science, and social equity.

The social positioning of Hispanos is altered through shifts in public priorities with respect to the management of national forest lands, through the globalization of the timber market, through the continued delegitimizing of the Hispano way of life and relationship with nature, and through Hispano engagement in forest management practice and policy processes. Since the mid-1980s, Hispano residents have sought to elevate their status in the timber industry from labourers to logging and wood products entrepreneurs. Just as they began pursuing this path, however, the globalization of the lumber market made lumbering in Vallecitos less and less profitable. And at the same time that this was occurring, environmental groups began demanding that greater care for the maintenance of ecological processes be taken in the management of national forest lands, and the Forest Service itself began to shift its orientation to ecosystem management. In addition, the devaluing of Hispano environmental credentials has continued unabated.

These developments began with an occurrence quite familiar in the history of the VFSYU—the designated operator asking the Forest Service to increase the allowable annual cut. Duke City Lumber Company officials argued that their sawmill at Vallecitos needed to process more timber to be economical. Forest Service officials eventually came around to this point of view, and in the draft management plan for the Carson National Forest released in 1985, they nearly doubled the unit's allowable annual cut. Local residents protested this development, as did environmental groups, and with the assistance of an environmental attorney who lived in the community at the time, they petitioned the Forest Service to include provisions in the plan that would encourage local entrepreneurship and community cooperatives. With a sympathetic forest supervisor, John Bedell, and a similarly sympathetic district ranger for the Vallecitos area, Gilbert Vigil, in office, the Forest Service accommodated the residents' wishes in a new policy for the unit that was written into the final forest plan. The new policy did increase the allowable cut, to a level that was only slightly less than originally proposed, but it set aside 1.0 million board feet of saw timber and 1.1 million board feet of forest products (firewood and other wood products) specifically to provide harvesting opportunities to small operators. The remaining 5.5 million board feet was available to Duke City Lumber Company on a non-competitive basis, as before (USDA, Forest Service 1986).

Having succeeded in getting provisions for small sales written into the prescriptions for the unit, the residents of the area set about positioning themselves to take advantage of them. In 1988 they formed a wood products cooperative called the Madera Forest Products Association (MFPA), and in 1990 they founded La Compañía Ocho logging company. The road has been rocky for these two entities, however. The MFPA, for example, has been stymied by a lack of equipment, buildings, and organizational capability among its members. In addition, the practice of offering firewood sales competitively has undermined the intended cooperative nature of the association. The association's own members bid against each other on such sales.

La Compañía, for its part, had difficulty securing its designation as a responsible operator for the unit and being offered small timber sales. In 1990 Forest Supervisor Bedell, who had helped La Compañía secure loans for equipment purchases by assuring the banks that small timber sales would be available to the company, was transferred to another national forest. After his departure, the Forest Service's approach to La Compañía changed dramatically. When the company requested approved responsible operator status in late 1990, it was told that it could not be so designated because it did not own a sawmill in the Vallecitos area. This decision appeared arbitrary, however, since the Duke City Lumber Company had been designated an approved responsible operator in 1972 but had not actually owned the sawmill at Vallecitos until 1990. The company had operated on the unit for 18 years, contracting with the owner of the mill to do the rough manufacturing of lumber required by the SYFMA. Members of La Compañía wondered why they could not do the same.

La Compañía and the MFPA acted on their dissatisfaction and filed a racial discrimination lawsuit against the Forest Service in 1994. Among other things, the lawsuit charged that the offering of fuel wood and other wood products for sale on a competitive basis to the MFPA, that delays in granting new responsible operator status to La Compañía, and that delays in offering small timber sales were the result of a racist bias within the agency and were therefore discriminatory actions (*Complaint, CIV 94-317 JB/LS*, 1994).

The outcome of the lawsuit had far-reaching consequences for the VFSYU, as well as for the members of the MFPA and La Compañía. While the US magistrate presiding over the case ruled in November 1994 that La Compañía need not own a sawmill to purchase timber sales on the VFSYU (Easthouse 1994), in 1996 the Forest Service and the plaintiffs settled out of court before any further rulings were made. The terms of the settlement included awarding more than three-quarters of the timber from scheduled sales on the unit to La Compañía. The Duke City Lumber Company responded to this development by giving its sawmill to a newly created, locally run non-profit organization called Las Comunidades and withdrawing completely from operations on the VFSYU. It had long considered the mill inefficient, and in the opinion of company officials, the company's having access to less than 25 per cent of the unit's sale rendered it completely unprofitable. As a consequence of these actions, for the first time in the unit's history, Hispano residents were in the position of owning and operating all of the forest products enterprises operating on it.

Yet the picture was not as rosy as it might seem. Throughout the period they were struggling to establish the MFPA and La Compañía, Vallecitos-area residents had also been struggling to come to terms with issues of threatened and endangered species on the VFSYU. In 1990, the same year that La Compañía was seeking designation as an approved operator for the unit, New Mexico environmentalists, as well as the state Department of Game and Fish, began raising concerns about the destruction of the Mexican spotted owl (listed as a threatened species by the US Fish and Wildlife Service in 1993) and goshawk habitat in northern New Mexico. Although the VFSYU Association and several local and national environmental groups tried to establish a coalition that would work jointly to promote sustainable forestry and preservation of spotted owl habitat on the unit, the effort quickly dissolved when neither the environmental groups nor the local residents lived up to the expectations that they had of each other.

The situation worsened for the MFPA and La Compañía in the fall of 1995, when a federal judge in Phoenix placed an injunction on all logging in critical habitat for the Mexican spotted owl. This brought virtually all logging on national forest lands in New Mexico and Arizona, as well as firewood harvesting in the Carson National Forest, to a standstill. This injunction was challenged in the courts, lifted, reinstated, lifted, and reinstated again in a brutal roller-coaster ride for the members of the MFPA and La Compañía. By the time the matter was finally resolved, La Compañía and Madera Forest Products had lost much of their labour force. Needing to feed their families, many of the workers had returned to the pattern of working jobs in Española, Santa Fe, and other towns outside the Vallecitos area. Las Comunidades, the new owner of the sawmill previously owned by the Duke City Lumber Company, was having difficulty marketing its lumber, and moreover, the sawmill needed major refurbishing.

The irony of the situation is that just as Hispano residents were succeeding in moving from the labour force to higher positions in the production chain, owning and operating their own businesses and cooperatives, the globalization of timber markets was undermining the profitability of the timber industry in northern New Mexico, a powerful constituency (environmentalists) was demanding changes in national forest management, and the Forest Service itself was reassessing its own management practices. This is where contradictions in the sustainability concept become evident. Although the movement of Hispanos up the production chain would seem to promise a more equitable distribution of the benefits of the successful management of the sustained-yield unit, the legacy of sustained yield embodied in the provisions of the SYFMA, together with the approaches to sustainability that environmentalists and Forest Service officials have adopted, places Hispano residents in the structure of the VFSYU's management in much the same position they were in in the 1950s.

The SYFMA still requires rough manufacture of lumber at a mill near Vallecitos, and Forest Service officials still interpret the Act to mean that timber is the primary focus of commodity production on the VFSYU. While these circumstances led the founders of La Compañía to develop their capacity for timber harvesting and led the founders of Las Comunidades to acquire Duke City's old mill for milling rough-cut lumber, the timber market is such that meeting the economic objectives of the VFSYU is even more difficult today than it was in the 1950s. A case in point involves the requirement of rough manufacture. Las Comunidades cannot sell the rough-cut lumber it produces because sawmills with finishing capabilities find it more costly to finish the nonprofit's rough-cut lumber than to produce finished lumber from raw logs themselves. Las Comunidades, in turn, lacks the capital to refurbish its mill (which still uses 1970s technology) to produce finished lumber.

The environmentalists' approach to sustainability reinforces the Hispanos' positioning in the political economy of forest management in northern New Mexico. That their primary concern has been with the preservation of ecological values—what they refer to as the maintenance of integral ecosystem functions—is evident in the many lawsuits they have filed in an effort to halt logging and other extractive activities on public lands, as well as in their assertions that any solution to environmental problems must be science based. The privileging of science as the knowledge system needed to develop sustainable land-use systems devalues Hispano environmental knowledge and continues to deny Hispanos the discursive legitimacy to engage meaningfully in negotiations over the management of natural resources.[4]

While environmentalists in northern New Mexico are by no means of a single opinion, the views they hold, with a few notable exceptions, tend to devalue Hispano relationships with the land. Some openly suggest that Hispanos lack respect for 'the basic biology that underpins sustainability.' One environmentalist I interviewed dismissed Hispano activists as 'professional Hispanics', implying that their activism is more for show than it is based on any real desire to address the needs of Hispano communities. Even environmentalists sympathetic to Hispano causes may unwittingly privilege Anglo approaches to conservation, suggesting, as one I interviewed did, that 'for some reason' Anglos have been the ones doing all the conservation. And those who express support for logging alternatives that could help Hispano enterprises may be chastised by the leaders of their organization. For example, one Anglo member of the Santa Fe group of the Sierra Club was reprimanded by regional and national officers of the club for supporting a logging alternative in a draft environmental impact statement for a timber sale on the VFSYU (Bird 1999).

Forest Service and other land management agency officials similarly doubt the motives and competency of Hispano residents of the Vallecitos area. Some argue that the sustained-yield unit is a social welfare program and that the local residents are just looking for a handout. Agency officials also argue that large industry is needed in the area to achieve management objectives. In addition, in my interviews with agency officials, I frequently encountered the sentiment that Regional Forester Pooler expressed in 1943—that Hispanos are clinging to an outmoded way of life and should get out of their communities and go where the jobs are—indicating that this view is still widespread in the agency. Weber (1991), of the US Minerals Management Service, similarly argues that Hispano culture is not suited to sustainable agricultural production and, therefore, that Hispanos would be better off joining the labour force elsewhere.

It is quite likely that the environmentalists and Forest Service officials I interviewed do not harbour any racial animus towards Hispanos. Their approach to forest management, which incorporates the representations of the Hispano relationship with nature I just described, however, has racializing effects. Even if they do not intend to discriminate, the devaluing of the Hispano way of life and relationship with nature that their discourse accomplishes contributes to maintaining the marginal social position of Hispanos, as well as to delegitimizing Hispano discourses.

In addition to devaluing the Hispano way of life, Forest Service officials on the national level have exercised little leadership in integrating equity into the agency's approach to sustainability. In response to changing public demands, to new understandings in ecological science, and to the catastrophic wildfires of 2000 and 2002, the agency now places greater emphasis on reducing wildfire fuels, conserving wildlife habitat, improving water quality, creating recreational opportunities, and preserving biodiversity.

CONCLUSION

Despite the Forest Service's original paternalistic approach to integrating ecological, economic, and social goals in the VFSYU's management, the unit did provide some benefits to Hispano residents of the Vallecitos area. The wages workers earned no doubt enabled many families to remain in the area. The availability of logging and milling jobs on the unit obviated the need to move to Albuquerque, Los Angeles, or elsewhere in search of work as so many families throughout the region are forced to do. In addition, the SYFMA of 1944 provided Hispanos with a legal resource they could use to support their claims to having a right to a say in the unit's management. By appealing to the community support provisions in the law, Hispano residents were also able to pressure the Forest Service to keep the VFSYU operating (other sustained-yield units have been shut down) and to change policy in ways that might bring them even greater benefits.

Nevertheless, the structuring of management, the delegitimizing of the Hispano relationship with nature, and changes in the global economy have served to maintain the racial order within the region, even in the absence of racial animus or intent to discriminate, and even in the context of calls to integrate economic, ecological, and sociological approaches to achieving sustainability. Whereas the structuring of management under sustained yield provided strategic advantages for large lumbering firms, the shifts in Forest Service management objectives, as well as in environmental groups' approaches to sustainability, have denied Hispano residents any strategic advantage in negotiating the terms of forest management, despite their current position as designated operators on the unit. Their experiential knowledge of the environment is not recognized as legitimate for informing management in any meaningful way, and the unit's management is still rooted in an environmental/racial discourse that devalues the Hispano way of life and relationship with nature. Hispano residents of the Vallecitos area are thus in much the same position in relation to the VFSYU's management as they were in the 1950s. Many environmentalists and people in the public at large do not consider that the Hispanos' expertise in logging and milling lumber can serve the cause of sustainability, and even if it did, the marketing of rough-cut lumber in the globalized timber market is problematic. The case of the VFSYU suggests, therefore, that current practice in integrating economics, ecological science, and social equity in natural resource management may be constituted of processes and discourses that reproduce the racial order. Without explicit attention to social positioning and to creating opportunities for, and conditions conducive to, the advancement of socially marginalized groups, the integration of issues of equity with ecological and economic concerns will remain an elusive goal.

NOTES

1. Unless otherwise noted, the historical documents cited pertaining to the VFSYU are on file at the El Rito Ranger District of the Carson National Forest, El Rito, New Mexico.
2. Forest Service officials had originally envisioned an excelsior plant, a facility for manufacturing box shook to supply fruit growers in Colorado's nearby San Luis Valley, and Christmas tree production as additional economic development activities associated with the VFSYU.
3. Local residents Albert Trujillo and Natividad Varela were the president and vice-president, respectively, of the union. I interviewed both in 1994.

4. While Hispanos may not have a perfect environmental record, neither does science. This has prompted Fortmann (forthcoming) and others to argue for the development of an integrated science that incorporates both scientific knowledge and the experiential knowledge of local peoples.

REFERENCES

Barrera, M. 1979. *Race and class in the Southwest: A theory of racial inequality.* Notre Dame: University of Notre Dame Press.

Bird, B. 1999. Statement of Sierra Club forest management policy violations in Grossman letter. On file in the possession of the author.

C.L.F. 1943. Memorandum to Earl H. Clapp, 21 September. Vallecitos File 2410.

Carlson, A. 1990. *The Spanish-American homeland: Four centuries in New Mexico's Rio Arriba.* Baltimore, MD: Johns Hopkins University Press.

Clary, D. 1985. *Timber and the Forest Service.* Lawrence, KS: University Press of Kansas.

Complaint, *CIV 94-317 JB/LS.* 1994. US District Court for the District of New Mexico.

Cooperrider, C. K., and B.A. Hendricks. 1937. Soil erosion and stream flow on range and forest lands of the Upper Rio Grande Watershed in relation to land resources and human welfare. *Rep. Technical Bulletin* 567 (USDA, Washington, DC).

Cottam, L.F. 1946. Memorandum for files, 4 June. Vallecitos File 2410.

deBuys, W. 1985. *Enchantment and exploitation: The life and hard times of a New Mexico mountain range.* Albuquerque: University of New Mexico Press.

Deutsch, S. 1987. *No separate refuge: Culture, class, and gender on an Anglo-Hispanic frontier in the American Southwest, 1880–1940.* New York: Oxford University Press.

Easthouse, K. 1994. Ruling favors Hispanic loggers. New Mexican, 13 December, B-4.

Ebright, M. 1994. *Land grants and lawsuits in northern New Mexico.* Albuquerque: University of New Mexico Press.

Fortmann, L. (ed.). Forthcoming. *Doing science together: The practice and politics of participatory research.* London: Blackwell.

Gross, L.S. 1953. Memorandum to the Record, 2 February. Vallecitos File 2410.

Jackson, J.W. 1952. Letter to Carson Forest Supervisor, 12 September. VFSYU File 2410.

Lindh, C.O. 1950. Letter to Senator Dennis Chavez, 25 July. Vallecitos File 2410.

——— 1952. Letter to Chief of the Forest Service, 17 September. Vallecitos File 2410.

Mason, D.T. 1927. Sustained yield and American forest problems. *Journal of Forestry* 25 (6):625–58.

Pooler, F.W. 1943. Letter to Associate Chief, Earle H. Clapp, 23 February. Vallecitos File 2410.

Rothman, H. 1989. Cultural and environmental change on the Pajarito Plateau. *New Mexico Historical Review* 64 (2): 185–211.

Schallau, C.H., and R.M. Alston. 1987. The commitment to community stability: A policy or shibboleth? *Environmental Law* 17:429–81.

Scott, D.O. 1947. Sustained yield case study. Vallecitos File 2410.

Sustained Yield Forest Management Act. 1944. In *58 Stat. 132, as amended; 16 U.S.C. 583-583i.*

USDA, Forest Service. 1947. Public hearing of Vallecitos Federal Sustained Yield Unit, 9 December. VFSYU File 2410.

——— 1986. *Carson National Forest Plan.* Washington, DC: USDA.

USDA, Soil Conservation Service. 1939. *Tewa Basin Study.* Albuquerque: Region 8, Division of Economic Surveys.

Wall, L.A., J.A. Egan. 1952. Management plan for the Vallecitos Federal Sustained Yield Unit. USDA, Forest Service, Carson National Forest. VFSYU File 2410.

Weber, K.R. 1991. Necessary but insufficient: Land, water, and economic development in Hispanic southern Colorado. *Journal of Ethnic Studies* 19 (2): 127–42.

Westphall, V. 1983. *Mercedes Reales: Hispanic land grants of the Upper Rio Grande region.* Albuquerque: University of New Mexico Press.

CHAPTER 12

Information Management for Water Resources: Concepts and Practice

Sarah Michaels, Dan McCarthy, and Nancy P. Goucher

INTRODUCTION

Generating, processing, and transmitting information have become so central to the organization of society that Castells (2000) has dubbed ours the information age lived in by informational society. This general state has permeated the environmental policy domain, where the importance of information has garnered increasing attention from practitioners and academics (Hornbeek 2000).

Information sharing, ecosystem management, accountability, and consensus are the four strategies agreed to by the Province of Ontario, the eight American states that border the Great Lakes, the two inter-tribal agencies, and the three federal agencies that signed the 1997 revision of *A Joint Strategic Plan for Management of Great Lakes Fisheries* (GLFC 1997). Each signatory agency is committed to establishing common standards for collecting, accessing, analysing, and sharing data (GLFC 2000).

Out of the necessity to address the problems associated with a complex, large ecosystem that crosses an international border, the Great Lakes Basin has provided the relevant parties a setting to implement or at least experiment with emerging approaches to integrated natural resource management. For example, ecosystem management approaches in the Great Lakes emerged in response to complex environmental problems arising from a large lake system with small watersheds and one of the largest concentrations of people and industry in North America (Slocombe 2004).

To restore and protect the planet's largest body of surface freshwater, the governments of Canada and the United States signed in 1972 the Great Lakes Water Quality Agreement. The present agreement was created in 1978, was revised in 1983, and had new annexes added through a 1987 Protocol. Article 8 of the 1978 agreement requires the International Joint Commission (IJC) to assess progress and help governments achieve the agreement's goals by preparing a biennial report on the water quality of the Great Lakes. The IJC was established by the 1909 Boundary Waters Treaty, its purpose being to assist in preventing and to help in resolving disputes about the use and quality of boundary waters. In the most recent biennial report, issued in September 2004, the IJC (2004–5) challenged governments to consider 'whether the sum of their policies, programs and management efforts are sufficient to protect water quality from the impact of continued expansion of its major urban areas in the Great Lakes Basin.' The IJC suggested that the lake basin level was the appropriate scale to use in responding to this issue, with participation from municipal, state/provincial, and federal governments (IJC 2004). Such a wide-sweeping endeavour requires drawing on and integrating what is known about water quality in the watersheds that drain into the Great Lakes. An initial challenge is to understand the logic and outcomes of how that information has been marshalled.

This chapter presents a case study of an initiative by Ontario's conservation authorities to develop an information management strategy through a provincial review of the state of water information in Ontario. A conservation authority is a quasi-government organization that employs an integrated strategy to manage a watershed on behalf of constituent municipalities. Watershed-scale water management is key to improving water management nationally and internationally. The information management initiative described here is important because it has the potential to enlighten decision making from the local scale through to the scale of the Great Lakes Basin and to build institutional capacity within conservation authorities.

The information management practices of conservation authorities should also be examined because of the authorities' heavy reliance on information-based policy instruments to achieve their mission. The authorities' organizational structure, goals, and forms reflect their reliance on, and the desirability of maintaining, a cooperative working relationship with their constituent municipalities, the provincial government, and others they must work with to manage, conserve, and protect natural resources on an integrated watershed basis. They cannot readily or extensively turn to regulation or economic incentives to achieve their ends. Information development efforts to support information-based policy instruments likely emphasize quality control in obtaining data, the importance of understanding the needs and characteristics of those at whom the policy instrument is directed, and the importance of effectively communicating the information (Hornbeek 2000).

The case study presented in this chapter deals with conservation authorities developing an information management strategy, and is situated in the literature on information management and knowledge management. With few exceptions, such as Simard 2000, little work has been on the direct applicability of information or knowledge management to natural resource management. In the next section, we provide an overview of data, information, and knowledge and then discuss information and knowledge management. This treatment is essential because the understanding of these concepts has become so blurred that these terms are on the verge of losing their explanatory value. We then present the case study, following this with reflections on theory-generated and practice-derived information management.

DATA, INFORMATION, AND KNOWLEDGE

Historically, data and knowledge have been understood as different aspects of information, the oldest of the three terms (Bawden 2001). Data results from measuring variables, such as voltages (Rouse 2002). Information is data put in context (Bouthillier and Shearer 2002). It is data collected in a comprehensive form that can be communicated and used, such as tables of statistics. 'What' questions generally are answered by information. Knowledge is information that humans have evaluated and organized and that can be used purposively, such as in explanations (Rouse 2002). Knowledge can be used to guide action and is predictive (Bouthillier and Shearer 2002). 'How' and 'why' questions are usually answered by knowledge (Rouse 2002). Attempts to articulate the relationship between data, information, and knowledge go back to the seventeenth century, shortly after the last of the three terms, data, appeared (Bawden 2001).

The distinction between data, information, and knowledge is well made in an example used by Bouthillier and Shearer (2002). If 10 degrees is an example of *data*, then *information* would be that it is 10 degrees outside and *knowledge* would be that 10 degrees is cold and requires dressing warmly. In a process-oriented systems approach, information and data are separate and linked, with information being produced through the combination of data and meaning (Bawden 2001). Nonaka and Takeuchi (1995) regard information as a stream of messages and knowledge as a creation of that stream, anchored in what stakeholders commit to and believe.

Miller (2002) argues that information becomes knowledge when it is interpreted by a human. Bawden (2001) suggests that one way to understand knowledge is to see it as a form of information that only exists within the mind of an individual. To be able to communicate it requires making it objective and recordable. The term for this 'objectivized' state is *information* (Bawden 2001). Analysing information may be a means to create knowledge (Bouthillier and Shearer 2002). We constantly transform information into knowledge and knowledge into information. All human learning and communication is based on the transformation of these separate and interacting entities (Orna 1999).

The utility and meaning of information are not dependent on collective use or sharing. So, from an organizational perspective, the individual consumption and use of information could be highly effective. In contrast, while knowledge may be acquired individually, to be useful, knowledge must be shared by a community of practice. For example, rules and procedures known only by one individual would be useless and meaningless (Bouthillier and Shearer 2002).

Pragmatically, what makes information and knowledge valuable is the extent to which they help users with their near-term plans, reduce uncertainty for users, and are delivered in an easily understood and usable form. Value then is user specific (Rouse 2002).

INFORMATION MANAGEMENT

The intent of managing information is to make sure that an organization has the information-based capabilities to adapt on an ongoing basis to evolving internal and external environments (Kirk 1999). Information management focuses on controlling, preserving, and retaining information (Bouthillier and Shearer 2002), and it can do this because information can be captured and stored in databases (Rouse 2002). Information management involves categorizing and systematizing information to make it trackable and accessible (Simard 2000). Information management initiatives focus more on systematically controlling recorded information than on using the records (Cronin 1985). Information management is about managing an organization's information resources by managing information technology (Wilson 1989; Bouthillier and Shearer 2002).

Information management, as we think of it now, came of age in the third quarter of the twentieth century. It was a time of unprecedented interest in information as a problem, accelerated technological development (e.g., in communications and information-processing technology), dramatic increase in the volume of scientific literature, and growing demand for new systems of organizing, storing, and retrieving information from huge US governmental research programs. In the late 1960s, information science emerged to develop new procedures and organizational arrangements to facilitate access to, and new ways of manipulating, recorded information (Rayward 1983).

Information management became popular as a term applicable to handling a particular subset of information characterized by being formal and structured and having a defined life cycle. This type of information is prevalent and important within business and government organizations and is reasonably amenable to computer processing. As a result, information management was enthusiastically adopted. Only gradually did it become apparent that much of the information that was most valuable to an organization, particularly that which enables creativity and innovation, was excluded. Recognition of this shortcoming led to the evolution of knowledge management as a means to deal with the process-oriented dimensions of information and then the more tacit form of information (Bawden 2001).

KNOWLEDGE MANAGEMENT

Knowledge management is about incorporating into the collective knowledge base individual and institutional learning to enable its utilization (Bukowitz and Williams 1999). The focus of knowledge management is on sharing knowledge (Bouthillier and Shearer 2002; Martensson 2000) because knowledge is available only from those who know (Rouse 2002). Knowledge management strategies refer to the means that organizations use to convert information into knowledge and use it to achieve their aims (Hovland 2003). In the corporate sector, the aim of knowledge management is to get to the next important discovery first (McElroy 2000). In the public sector, knowledge management is used to improve operations (Al-Hawamdeh 2002) and enhance service performance (Al-Hawamdeh 2002; Rocheleau 2000).

The Two Generations of Knowledge Management

First-generation knowledge management is about enabling people to share what they know explicitly by capturing, codifying, and distributing their knowledge. It is about enabling people to do their jobs more effectively by 'getting the right information to the right people at the right time' (McElroy 2003, xxiv). Its supply-side orientation involving codification and distribution promotes the intentional imitation of best practices (McElroy 2000).

The first generation of knowledge management has been discredited for providing solutions that are essentially the repackaging of information retrieval systems. It has been dismissed as '"little more than a re-hash" of yesterday's "information management" schemes' (McElroy 2000, 199).

Second-generation knowledge management is about helping people articulate and share what they know implicitly. The emphasis is on producing and sustaining the conditions necessary for producing knowledge (McElroy 2003) by developing an 'implementation strategy for organizational knowledge creation and learning' (McElroy 2000, 200). Its demand-side orientation promotes producing and sustaining conditions for producing knowledge (McElroy 2000).

Hovland (2003) points out that an organization's ability to learn can be divided into first- and second-order strategies (Argyris 1992) that correspond with first- and second-generation knowledge management strategies. According to Argyris (1992), first-order strategies involve 'single-loop learning' and aim to get the job done by correcting errors and taking action that fits within existing policy. In contrast, second-order strategies involve 'double-loop learning' and aim to increase an organization's capacity to think critically about policy frameworks. Within the context of first-generation knowledge management, single-loop learning reinforces the concern with efficiency by enhancing day-to-day processes (McElroy 2000). Within the context of second-generation knowledge management, the concern with efficacy is reinforced.

Distinguishing between Information Management and the Two Generations of Knowledge Management

Table 12.1 presents some of the key dimensions along which information management and first- and second-generation knowledge management are distinguished. While there has been a tendency to treat the spectrum from information management to second-generation knowledge management as an evolutionary trajectory (McElroy 2000, 2003; Martensson 2000), we argue that it is more constructive to view each as complementary to the other. The choice of approach should be a function of the intended outcome of the activity. The case study that follows is not a rudimentary effort at knowledge management. Rather it is an aptly framed information management initiative.

CASE STUDY: ONTARIO'S CONSERVATION AUTHORITIES APPROACH INFORMATION MANAGEMENT

Ninety per cent of the people in Canada's most populous province live in watersheds in which conservation authorities function (Conservation Ontario 2000a). The conservation authorities of Ontario operate in the environmental policy domain as community-based, quasi-government agencies to conserve, restore, develop, and manage natural resources through an integrated approach, on a watershed basis. They work in a policy milieu in which no single entity on its own has all the knowledge or resources needed to execute policy (Rhodes 1996). Consequently, to fulfil their obligations, they must continually engage with government, groups, and individuals in

TABLE 12.1 Comparing information management and first- and second-generation knowledge management

	Information Management	Knowledge Management Generation I	Knowledge Management Generation II
Purpose	Systematically controlling recorded information	Dissemination	Education Innovation
What Is Being Managed	Formal, structured information with defined life cycle	Conditions for sharing explicit knowledge	Conditions for knowing and for sharing explicit and explicit knowledge
Organizational Learning	Single loop	Single loop	Double loop
Target Scale	Individual	Individual member of community practice	Collective membership of community practice
Activities Encouraged	Consumption Use	Sharing	Creation
Promotes	Replicatioon	Imitation	Innovation
Driven By	Supply of accessible informaton	Supply of existing information and knowledge	Demand for new organizational knowledge
Primary Organizational Concern Addressed	Efficiency	Efficiency	Efficacy

SOURCES: Cronin 1985; Bouthillier and Shearer 2002; McElroy 2000; Hovland 2003.

'social-political governance' (Kooiman 1993, 2). The nature of the problems they address are complex, evolving, and multi-faceted, with no one player having the monopoly on the wherewithal to come up with independent, workable solutions (Kooiman 1993).

Thirty-eight conservation authorities were formed under the Conservation Authorities Act, 1946. On-the-ground initiatives of the 36 conservation authorities currently operating vary as a function of differences in physical geography, issues, resources, and expertise (Conservation Ontario 2001). They all plan, coordinate, and manage on behalf of municipalities within a watershed and work with all levels of government to coordinate and deliver programs and services that address water quality, water supply, flooding, natural areas protection, public education, and recreation (Conservation Ontario 2000a).

Flooding, drought, degraded water quality, and associated problems in Southern Ontario from the early twentieth century on prompted the passage of the Conservation Authorities Act, 1946 (Conservation Ontario 2001). The conceptual basis for the conservation authorities goes back to integrated resource management experiences in New Zealand, England, and Wales. At the same time, they build on the legacy of innovative integrated water resource management that developed through the Ohio Conservation Districts and the Tennessee Valley Authority in the United States (Mitchell and Shrubsole 1992).

To evaluate and improve the state of water information in Ontario, the Water Resource Information Project (WRIP) was initiated early in 2000. The intent of the project was to review Ontario's state of water information, strategically focusing on 'identifying problems with current water information' (Wilcox 2001, 21) and developing 'a strategy to ensure information is readily available to support effective water management' (Wilcox 2001, 6). The WRIP is a cooperative effort involving the Provincial Land and Resources Cluster of Ministries (PLRCM) and Conservation Ontario (Wilcox 2002, 4). At the time, the PLRCM included Agriculture and Food; Environment and Energy; Natural Resources; and Northern Development and Mines. The intent of creating a cluster was to provide 'a common and integrated approach that will lead to more effective and efficient delivery of information and information technology (I&IT) solutions' (Office of the Corporate Chief Information Officer 2002). Conservation Ontario is a non-governmental umbrella organization formed by the conservation authorities in the late 1970s to represent their interests. On behalf of its constituent conservation authorities, Conservation Ontario's mission is to 'raise awareness, build relationships and influence decision-makers' (Conservation Ontario 2000b).

This discussion of WRIP focuses on selected aspects of what Conservation Ontario set out to do through its WRIP work. Conservation Ontario was invited to partner with the Ontario Ministry of Natural Resources (MNR) to develop WRIP to ensure that the project addressed conservation authority (CA) water business functions (Wilcox 2002). In the first year of WRIP, through literature reviews, interviews with conservation authority general managers and senior staff, and workshops attended by 75 technical staff, six 'water business functions' were identified including:

- protect life and property from flood erosion
- encourage a sustainable water supply
- inventory and monitor water quality
- assess and report on water conditions
- protect/enhance water quality
- provide recreational/quality of life opportunities' (Wilcox 2002, 4)

Conservation Ontario used the opportunity of WRIP to articulate a linear, five-element approach to information management. See Table 12.2.

Developing the conservation authorities' model of information management served four purposes.

Table 12.2 Components of conservation authorities' perspective on information management

Component	Description	Example
Design	'...purpose and types of decision to be supported by the information gathering system' (Wilcox 2002, 6)	'...water monitoring network design, where the system is developed to answer appopriate water management questions' (Wilcox 2003, 6)
Collection	'...method of physically gathering the data' (Wilcox 2002, 6)	'...field water sampling or water level measurement through telemetry, who collects it, what standards are used' (Wilcox 2003, 6)
Storage	'...method of storing the data ...affects access and sharing of data' (Wilcox 2002, 6)	'...databases, spreadsheets, paper files, standards' (Wilcox 2003, 6)
Analysis	'...method of converting data to information for decision making' (Wilcox 2002, 6)	'...hydrologic models, trends over time, comparison to predetermined thresholds' (Wilcox 2002, 6)
Reporting	'...method of communicating information to decision makers and stakeholders' (Wilcox 2002, 6)	'...state of the watershed reorts, directors reports, etc.' (Wilcox 2003, 6)

SOURCES: Wilcox 2002, 2003.

1. It was a means to scope WRIP activities.
2. It served as a diagnostic tool to identify water-related information management weaknesses.
3. It acted as an aid to move strategically towards solutions.
4. It helped to broaden the appreciation of what constitutes information management beyond a narrow concern about data storage.

The formal responsibility for data storage rests in most cases with the provincial or federal government (Wilcox 2003).

Developing the Conservation Authority Information Management Strategy was a deliberate exercise in coming up with a plan to improve conservation authorities' capacity to manage information and nudge the provincial and federal governments to develop broader standards that could be used across jurisdictions (Wilcox 2003). Conservation Ontario (2001) recognized the role of provincial and federal government in establishing and enforcing water resources management regulatory standards, provincial scale monitoring, and science and technology transfer. The two-pronged approach of developing internal capacity and encouraging concomitant capacity building in partners is illustrated by the 2003 *Conservation Ontario Discussion Paper: Recommendations for Monitoring Ontario's Water Quality*. The paper's intent was twofold. First, it was intended to serve as a technical guide for conservation authorities in the process of developing their own water-quality monitoring systems. Second, it was meant to be 'a marketing/lobbying tool' to stimulate discussion about improving monitoring in the province with municipal, provincial, and federal partners (Wilcox 2003, 10). That a marketing plan was required to do the latter indicates recognition that, for implementation, it is necessary, but not sufficient, to have a good idea about what to do.

Conservation authorities and their association, Conservation Ontario, debate the extent and means of cooperation among themselves. This issue has taken on particular salience since the mid-1990s, when the provincial government largely withdrew from the conservation authority arena (Kreutzwiser 1998). On the one hand, conservation authorities do share the fundamental mission as set out in the 1946 legislation that governs them. On the other hand, they are each distinctive, independent operators, responsive to their constituent jurisdictions, and their institutional emphases are functions of local challenges and accessible resources. While all conservation authorities perform similar tasks in managing information, there has been little pooling of experience or expertise to aid the advance of the collective (Wilcox 2002). Under the circumstances, it is not surprising that three recommendations from the third year of WRIP listed in *WRIP III Conservation Ontario report* (Wilcox 2003)—facilitating regional database training for CA staff, identifying and sharing CA information technology expertise, and facilitating regional technical forms in aid of the sharing of information and expertise—proved challenging. At the technical level, there was no formal precedent or mechanism for sharing resources, expertise, and training. The acceptable solution was not to develop a formal structure to address these three, but rather to develop informal, individual solutions. No software application standard was identified. Minimum-standard database training was to be sought by individual CAs through educational institutions and private business rather than through a specific training program for CAs developed by their association. At the same time, the first formal training that CA staff received in sampling for the Provincial Water Quality Monitoring Network was designed and delivered by Conservation Ontario, the Upper Thames River Conservation Authority, and the Saugeen Valley Conservation Authority working with the Environmental Monitoring and Reporting Branch (EMRB) of the Ministry of Environment (MOE) (Wilcox 2003). This initiative was particularly salient because during WRIP's initial consultation with conservation authorities in 2000, improving Ontario's water monitoring networks was identified as the number one requirement for improving water information (Wilcox 2002). The problem stems from the lack of baseline water monitoring to support decision making in Ontario effectively. The need is for baseline monitoring solutions that are long term and scale appropriate for managing watersheds (Wilcox 2001).

Much of the current policy and institutional approach to water management in Ontario is a function of governmental and societal reaction to the water crisis in

Walkerton, a small town in southern Ontario. In May 2000 Walkerton's drinking water was found to be contaminated with deadly bacteria, primarily *Escherichia coli* O157:H7. Seven people died and 2,300 became ill. In direct response to this event, the Government of Ontario held an inquiry into the safety of drinking water across the province (O'Connor 2002). Concerns about the management of water information and other water-related issues have increasingly been influenced by the institutional intent to reduce the likelihood of a Walkerton-like incident. While the first three years of WRIP activity were consistent with the intent of source water protection planning, in 2003/4 WRIP IV began to make water information improvements under a more formal umbrella of source protection planning (Wilcox 2003).

The transition to improving water information within a more formalized, programmatic regime shifts to some extent the context for making water information improvements. Hornbeek (2000) suggests that environmental agencies develop information to serve two broad purposes. First, information can be used to direct or justify particular program activities. Using risk and exposure estimates to establish maximum contaminant levels for drinking water is an example of technical analysis designed for program support. Program support information is directed towards a defined target audience with the intent of guiding particular behaviour. Second, information can be used to benchmark or assess trends associated with environmental conditions or specific activities that affect the environment. An example of gathering benchmark information is collecting ambient water-quality information to determine how clean the water supply being tested is. Future WRIP activity may be more geared to the first purpose than it has been in the past. Historically, conservation authorities have emphasized the value of benchmarking information for a broad-based audience of officials, the public, and stakeholders.

REFLECTIONS ON THEORY-GENERATED AND PRACTICE-DERIVED INFORMATION MANAGEMENT

The process model of information management proposed by Choo (1998) is not radically dissimilar from the organically developed approach developed by Ontario conservation authorities. Choo (2000) favours information systems designed to aid in solving work-related problems and to deal with specific problems by providing people with useful information. The conservation authorities recognize that to ensure that a system of information management is effective, there must be trained staff, provincial standards, and appropriate tools for each component (Wilcox 2002). While Choo's (1998) model is explicitly cyclical, the conservation authorities' approach is not. Each of Choo's steps requires planning, organizing, coordinating, and controlling activities that are supported by information technology. Information technology underpins the conservation authorities' approach. The parallels between the two are most striking when the steps of one are compared with the components of the other. See Table 12.3. Information is understood as a commodity that can be processed through an information production chain (Braman 1989). Choo's steps and the CAs' components are elements in their respective information production chains.

TABLE 12.3 Comparing steps in Choo's (1998) process model of information management with the components of conservation authorities' perspective on information management (Wilcox 2002)

Choo's Steps in Processing Information	CAs' Components of Information Management
Identify needs	Design
Acquire	Collect
Organize	Store
Distribute	Analyse
Use	Report

SOURCES: Choo 1998; Wilcox 2002.

The information production chain perspective brings to the fore the distinction and relations between elements (Braman 1989). Sensitivity to the distinction and relations between elements comes to the fore when different entities are responsible for different elements. For example, the conservation authorities are primarily responsible for collecting water-quality information, while the province takes the lead in storing it.

The notion of information as a commodity includes the exchange of information among people, the activities associated with that exchange, and the use of information. The social structure is complex. However, the complementary concept of a production chain for information enables policy-makers at one or a number of stages in the chain to exclude certain types of information, actors, and actions (Braman 1989).

To move beyond the conservation authorities' current conceptualization of information management will involve incorporating other understandings of information from Braman's (1989) hierarchy in addition to information as commodity. Information as perception of pattern adds context, recognizes that information has a past and a future, is influenced by motive and environmental and causal factors, and also has effects itself. This relativistic approach requires an explicit statement of the point of view from which it is applied. Since perception of pattern and context shifts from observer to observer, it is difficult to use the perception of pattern class of information definitions in policy-making.

Information is granted an active role in shaping its context when it is defined as a constitutive force in society. Information creates the social structure while being embedded in it. A particular vision of how society ought to be is supported by every information policy decision. Consequently, when information is defined as a constitutive force in society, making any policy decision about information involves deciding how society is to be structured. As a result, this definition of information is most effectively used at the outset of decision making as the first level of analysis and as the standard for evaluating decisions that treat information as a commodity. This type of definition may be used at the outset to expand the context in which other definitions are employed. In subsequent stages, narrower operational definitions may be employed appropriately, particularly since it is difficult to quantify events and effects when approaching information as a constitutive force in society (Braman 1989).

Complementary to Braman's (1989) hierarchy are applicable insights from complex systems. Implicit in the conservation authorities' approach to information management are elements of a systems approach. A complex systems perspective has led to an understanding of ecological and human institutional systems as complex, highly interconnected, perspective and context dependent, and rife with uncertainty (Holling 1995, 2001; Berkes and Folke 1998; Kay et al. 1999). A critical systems thinking approach (Jackson 2000; Midgley 2000) views information and knowledge for decision making as complex, contingent, and open to critical reflection.

The articulation of the components of information management by the conservation authorities can be seen as an initial cut at a systems description. The institutional context was acknowledged in the recognition of both the need for consistent information management practices and the CAs' varying capacities to achieve this. Perspective dependence was acknowledged and the value of multiple perspectives recognized by the bringing to bear of perspectives from all conservation authorities to develop the Conservation Authority Information Management Strategy (Wilcox 2001). Reviewing the state of water information in Ontario can be seen as an exercise in reflecting critically on the practices of the time.

Explicit incorporation of some ideas from critical systems thinking may strengthen the resilience of the conservation authorities' approach to information management. Here are two possibilities. The first is to develop a complex systems description of the information and knowledge needs for decision making. This involves combining a description of what is currently known to be needed for decision making with needs that may arise owing to expected or unexpected change. This could include describing the types of information and knowledge required during normal operations, operations under stress (known or anticipated change), and operations under fundamental change (unanticipated or contextual change). The second is to engage in critical reflection about the information management process. One element of this would be to make explicit the assumptions underpinning the data, information, and knowledge used for decision making. Such an exercise would aid users by clarifying the strengths and limitations of a critical dimension of decision making. The implementation of either of these possibilities requires an institutional ability and willingness to think beyond the immediate crisis and to disclose what insiders may prefer to leave unmentioned.

CONCLUSION

Fuller consideration of the applicability and utility of information management and knowledge management to natural resource management will require more explicit consideration of whether, and if so how, they mesh with the application of established strategies to natural resource management, such as integrated resource management (Mitchell 1998) or an ecosystem approach (Slocombe 2004). In the meantime, this chapter has presented the experience of conservation authorities in their development of an information management strategy that addresses their water business functions within the framework of a review of the state of water information in Ontario. To move towards integrated water resource management, conservation authorities will need to build upon what they have learned about information management through joint endeavours, such as WRIP, to build a knowledge strategy that is as flexible as their business and the context in which they function. The knowledge strategy must be integrated, holistic, evolutionary, and capable of supporting the sharing of not only data and information but also knowledge. A truly integrated knowledge strategy will also function across the boundaries that exist between each of the conservation authorities, between various water and non-water-related issues, between disciplines, and between people in different levels of organizational hierarchies.

The conservation authorities have developed an information management strategy within the context of the evolving fields of information management and knowledge management while drawing to a limited degree on complementary insights from systems thinking. The experience in practice of developing an information management strategy bears a striking resemblance to what scholars envision. While academics and consultants have pursued knowledge management frameworks as a higher and more developed form of information management, the conservation authorities have appropriately framed their initiative as what it is—an information management endeavour. Arguably, while knowledge management is more encompassing than information management, as the WRIP example illustrates, information management has a vital role to play in advancing integrated resource management. Developing sound information management practices is critical for making decisions about individual watersheds and every aggregated scale that builds on them. The initiative of Ontario conservation authorities provides one strand in the cable of inputs into integrated natural resources decision making about the Great Lakes Basin.

ACKNOWLEDGMENTS

The authors wish to acknowledge the financial support provided by Social Sciences and Humanities Research Council grant number 410-2002-1418 (Organizational Knowledge Creation for Watershed Management, Principal Investigator S. Michaels), and additional support from the Canadian Water Network. The insights of Ian Wilcox, general manager/secretary-treasurer, Upper Thames River Conservation Authority, were indispensable. The editors of this volume, Kevin Hanna and D. Scott Slocombe, and the anonymous reviewers imparted constructive comments on a previous draft. Earlier versions of this chapter were presented at the Canadian Water Network Policy and Governance Symposium, Renison College, University of Waterloo, Waterloo, Ontario, 8 January 2004; the International Association of Great Lakes Researchers'

Annual Conference on Great Lakes Research: Great Lakes Need Great Watersheds, University of Waterloo, Waterloo, Ontario, 24 May 2004; the International Symposium on Society and Resource Management 2004: Past and Future, Keystone Resort, Keystone, Colorado, 3 June 2004; and the International Society of Ecological Economics, 8th Biennial Scientific Conference, Montreal Convention Centre, Montreal, Quebec, 12 July 2004.

REFERENCES

Al-Hawamdeh, S. 2002. Knowledge management: Re-thinking information management and facing the challenge of managing tacit knowledge. *Information Research* 8 (1), paper no. 143. Retrieved 8 October 2004 from http://informationr.net.proxy.lib.uwaterloo.ca/ir/8-1/paper143.html.

Argyris, C. 1992. *Overcoming organizational defenses: Facilitating organizational learning.* Boston: Allyn and Bacon.

Bawden, D. 2001. The shifting terminologies of information. *Aslib Proceedings* 53 (3): 93–8.

Berkes, F., and C. Folke. 1998. *Linking social and ecological systems: Management practices and social mechanisms for building resilience.* New York: Cambridge University Press.

Bouthillier, F., and K. Shearer. 2002. Understanding knowledge management and information management: The need for an empirical perspective. *Information Research* 8 (1), paper no. 141. Retrieved 5 October 2004 from http://InformationR.net/ir/8-1/paper141.html.

Braman, S. 1989. Defining information: An approach for policymakers. *Telecommunications Policy* 13 (3): 233–42.

Bukowitz, W.R., and R.L. Williams. 1999. *The knowledge management fieldbook.* Harlow, England: Financial Times Prentice Hall.

Castells, M. 2000. *The rise of the network society.* Vol. 1 of *The information age: Economy, society and culture.* 2nd edn. Oxford: Blackwell.

Choo, C.W. 1998. *Information management for the intelligent organization: The art of scanning the environment.* Medford, NJ: Information Today.

—— 2000. Closing the cognitive gaps: How people process information. In D.A. Marchand and T.H. Davenport (eds), *Mastering information management*, 245–53. London: Financial Times Prentice Hall.

Conservation Ontario. 2000a. *Conservation Ontario corporate profile.* Retrieved 27 July 2004 from http://conservation-ontario.on.ca/profile/profile.htm.

—— 2000b. *Conservation Ontario frequently asked questions.* Retrieved 29 September 2004 from http://www.conservation-ontario.com/faqs/faqs.htm#5.

—— 2001. *The importance of watershed management in protecting Ontario's drinking water supplies.* Submitted to the Walkerton Inquiry 20 March 2001. Newmarket, ON: Conservation Ontario.

Cronin, B. 1985. Introduction to B. Cronin (ed.), *Information management: From strategies to action,* vii–ix. London: Aslib.

GLFC (Great Lakes Fisheries Commission). 1997. *A Joint Strategic Plan for Management of Great Lakes Fisheries.* Ann Arbor, MI: Great Lakes Fisheries Commission.

—— 2000. *A Joint Strategic Plan for Management of Great Lakes Fisheries.* Fact Sheet 10. Ann Arbor, MI: Great Lakes Fisheries Commission.

Holling, C.S. 1995. What barriers? What bridges? In L.H. Gunderson, C.S. Holling, and S.S. Light (eds), *Barriers and bridges to the renewal of ecosystems and institutions*, 3–36. New York: Columbia University Press.

———— 2001. Understanding the complexity of economic, ecological, and social systems. *Ecosystems* 4:390–405.

Hornbeek, J. 2000. Information and environmental policy: A tale of two agencies. *Journal of Comparative Policy Analysis: Research and Practice* 2:145–87.

Hovland, I. 2003. *Knowledge management and organizational learning: An international development perspective. An annotated bibliography.* Working Paper 224. London: Overseas Development Institute.

IJC (International Joint Commission). 2004. *Twelfth biennial report on Great Lakes water quality.* Detroit: International Joint Commission.

Jackson, M.C. 2000. *Systems approaches to management.* New York: Kluwer.

Kay, J.J., H. Regier, M. Boyle, and G. Francis. 1999. An ecosystem approach for sustainability: Addressing the challenge of complexity. *Futures* 31 (7): 721–42.

Kirk, J. 1999. Information in organisations: Directions for information management. *Information Research* 4(3). Retrieved 8 October 2004 from http://informationr.net.proxy.lib.uwaterloo.ca/ir/4-3/paper57.html.

Kooiman, J. 1993. Social-political governance: Introduction. In J. Kooiman (ed.), *Modern governance: New government-society interactions*, 1–8. London: Sage.

Kreutzwiser, R. 1998. Water resources management: The changing landscape in Ontario. In R.D. Needham (ed.), *Coping with the world around us: Changing approaches to land use, resources and environment*, 135–48. Department of Geography Publication Series No. 50. Waterloo, ON: University of Waterloo.

McElroy, M.W. 2000. Integrating complexity theory, knowledge management and organizational learning. *Journal of Knowledge Management* 4 (3): 195–203.

———— 2003. *The new knowledge management: Complexity, learning and sustainable innovation.* New York: Butterworth Heinemann.

Martensson, M. 2000. A critical review of knowledge management as a management tool. *Journal of Knowledge Management* 4 (3): 204–16.

Midgley, G. 2000. *Systemic intervention: Philosophy, methodology, and practice.* New York: Kluwer Academic/Plenum Publishers.

Miller, F.J. 2002. I = 0 (Information has no intrinsic meaning). *Information Research* 8 (1), paper no. 140. Retrieved 8 October 2004 from http://informationr.net.proxy.lib.uwaterloo.ca/ir/8-1/paper140.html.

Mitchell, B. 1998. *Sustainability: A search for balance.* Waterloo, ON: University of Waterloo.

Mitchell, B., and D. Shrubsole. 1992. *Ontario conservation authorities: Myth and reality.* Department of Geography Publication Series No. 13. Waterloo, ON: University of Waterloo.

Nonaka, I., and H. Takeuchi. 1995. *The knowledge-creating company.* New York: Oxford University Press.

O'Connor, D.R. 2002. *Report of the Walkerton Inquiry: A strategy for safe drinking water. Part Two.* Toronto: Ontario Ministry of the Attorney General.

Office of the Corporate Chief Information Officer. 2002. *Land and resources cluster.* Retrieved 30 September 2004 from http://www.cio.gov.on.ca/scripts/index_.asp?action=31&P_ID=431&N_ID=1&P.

Orna, E. 1999. *Practical information policies.* 2nd edn. Aldershot: Gower.

Rayward, W.B. 1983. Library and information sciences: Disciplinary differentiation, competition, and convergence. In F. Machlup and U. Mansfield (eds), *The study of information: Interdisciplinary messages*, 343–63. New York: John Wiley and Sons.

Rhodes, R.A.W. 1996. The new governance: Governing without government. *Political Studies* 44 (3–5): 652–67.

Rocheleau, B. 2000. Prescriptions for public-sector information management. *American Review of Public Administration* 30 (4): 414–35.

Rouse, W.B. 2002. Need to know—Information, knowledge, and decision making. *IEEE Transactions on Systems, Man and Cybernetics—Part C: Applications and Reviews* 32 (4): 282–92.

Simard, A.J. 2000. *Managing knowledge at the Canadian forest service*. Ottawa: Science Branch, Canadian Forest Service, Natural Resources Canada.

Stinchcombe, A.L. 1990. *Information and organizations*. Berkeley: University of California Press.

Slocombe, D.S. 2004. Applying an ecosystem approach. In B. Mitchell (ed.), *Resource and environmental management in Canada*, 420–41. Don Mills, ON: Oxford University Press.

Wilcox, I. 2001. WRIP *water resources information project*. Conservation Ontario Report. Newmarket, ON: Conservation Ontario.

——— 2002. *Water resources information project Conservation Ontario report: 2001–2002. A Conservation Authority Information Management Strategy.* Newmarket, ON: Conservation Ontario.

——— 2003. *Water resources information project:* WRIP III *Conservation Ontario report.* Newmarket, ON: Conservation Ontario.

Wilson, T.D. 1989. Towards an information management curriculum. *Journal of Information Science* 15 (4–5): 203–10.

CHAPTER 13

Integrated Approaches for Transboundary Wildlife Management: Principles and Practice

Michael S. Quinn

INTRODUCTION

Wolverines (*Gulo gulo*), like other large carnivores, occupy large home ranges and possess a predilection for expansive dispersal. The first Global Positioning System (GPS) collar to be fitted on a North American wolverine was deployed early in 2002 in Wyoming's Grand Teton Mountains. The GPS collar technology allowed wolverine researchers in the Greater Yellowstone Ecosystem to better understand the habits and movements of this elusive creature. The pioneering wolverine lived up to its reputation for long-range dispersal. In a 42-day period, the animal travelled a minimum of 874 km (Inman et al. 2004). The winter travels of this sub-adult wolverine encompassed the rugged terrain of nine distinct mountain ranges, an area that included lands under the management jurisdiction of three states, two national parks, three national forests, one Bureau of Land Management (BLM) unit, and Bureau of Indian Affairs (BIA) lands. One of the poignant lessons we can learn from this story is that there is a need for integrated approaches to land and resource management that transcend jurisdictional boundaries.

Similarly, in the summer of 1991, a satellite collar was fitted to a five-year-old female wolf (*Canus lupus*) in Alberta's Kananaskis Country (Paquet 2005). The wolf, dubbed 'Pluie' by the researchers, remained in the area of her capture until late fall, when she initiated a series of travels that would eventually cover an area of over 100,000 km^2 in the four and a half years that she was collared (Figure 13.1). Pluie's extensive movements in the transboundary region of the Rocky Mountains encompassed more than 30 provincial, state, and federal jurisdictions.[1] The detailed information on the movements of such animals is essential to developing coherent management plans to ensure the long-term viability of populations and their habitats.

Land management approaches differ greatly across borders, resulting in habitat fragmentation and barriers to connectivity. In addition, as animals traverse borders and boundaries, they are subject to changing management regimes and associated ecological effects (Madonna 1995). For example, if Pluie were to pause on the boundary of Waterton Lakes National Park at the Canada/US border, a single step could

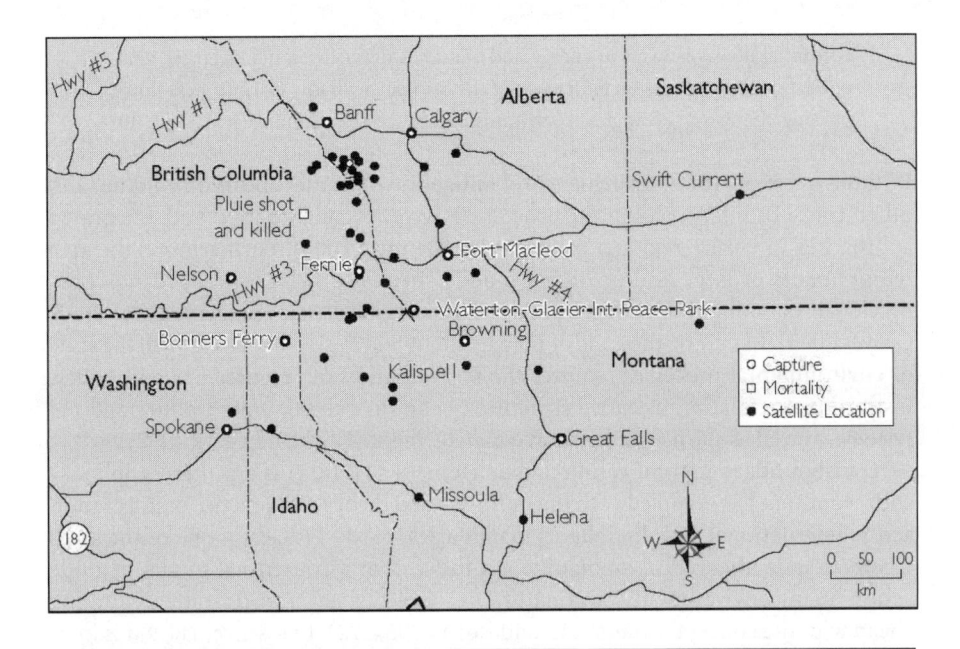

FIGURE 13.1 Satellite locations for female wolf 'Pluie', originally collared in Peter Lougheed Provincial Park, Alberta

take her from being protected under the US Endangered Species Act, to being federally protected in a Canadian national park, to being shot without restriction by a private landowner (Musiani and Paquet 2004).

The long-range, transboundary movements of the wolverine and wolf described above are impressive, but far from unique (see, for example, Boyd and Pletscher 1999; Merrill and Mech 2000; Vangen et al. 2001). Ecological processes such as dispersal and seasonal migration result in mammals, birds, and fish having to negotiate borders and boundaries around the globe. Westing (1993, 1998) estimates that approximately one-third of our planet's biodiversity 'hot spots' straddle national boundaries. The administrative lines and rectilinear grids that we have imposed on the landscape for historical, political, and socio-economic purposes rarely correspond to ecological spatial patterns (Forman and Moore 1992; Forman 1995; Chester 2006).

When borders include fences, they can have a direct negative effect on wildlife movement (Hobbs 1981; Griffin et al. 1999). However, even where boundaries are not marked by physical barriers, jurisdictional policy, planning, and management discontinuities have profound effects on ecological systems.

> An administrative border is like a glass wall that may not be readily apparent, but because nearly all terrestrial and aquatic ecosystems are open systems requiring

continual flows or fluxes of energy and matter, differences in management goals and land-use practices on either side of the border inevitably disrupt these flows, causing changes in ecological conditions and processes. (Landres et al. 1998, 40)

Without question, the management and mitigation of border effects on wildlife (and other transboundary ecological processes) require collaborative and integrated approaches to ensure regional connectivity (Bennett 2003). Furthermore, the area requirements of large and/or highly mobile species in heterogeneous landscapes necessitate planning and management at an (eco)regional scale (Sanderson et al. 2002; Groves 2003). Matching the scale of management and decision making with the scale of environmental processes requires the re-evaluation of resource law and policy, institutional and administrative structures, economic development policy and tax systems, as well as science and resource management (Glick and Clark 1998).

Transboundary natural resource management (TBNRM) is an interdisciplinary endeavour that has grappled with the multi-faceted theory and practice of integration across international and disciplinary boundaries. TBNRM is clearly one of the areas where, despite 'theoretical ambiguity and the lack of a consensual model to guide action, many environmental and natural resource agencies worldwide are forging ahead with the concept' (Margerum and Born 1995, 372). In essence, TBNRM is integrated management in practice.

The purpose of this chapter is twofold: first, to discuss the growing international movement towards TBNRM, and second, to provide a summary of critical lessons learned from the field of practice. The first part of the chapter provides a brief review of the current theory and practice in transboundary management and transboundary protected areas. The next section presents two case studies of TBNRM initiatives to illustrate the different approaches taken in North America and Europe. The case studies were carried out in regions that have the oldest formal transboundary initiatives in the world and mountain environments with comparable ecological assemblages. The first study concerns initiatives from the Crown of the Continent Ecosystem in the Rocky Mountains of North America, and the second, the Carpathian Ecoregional Planning Initiative in central Europe. Key principles and practices for integrated wildlife management in transboundary contexts will be summarized in the concluding section.

TRANSBOUNDARY NATURAL RESOURCE MANAGEMENT

The enterprise of integrated environmental management is, by definition, a boundary transcendence, not just of the administrative and jurisdictional lines that appear on maps, but also of the modernist divisions between the studies of nature and culture (Berkes and Folke 1998; Fall 2003). A convergence in the literature from such fields as conservation biology (Noss and Cooperrider 1994; Soulé and Terborgh 1999), ecosystem-based management (Grumbine 1994; Redmond 1999; Quinn 2002), regionalism and regional planning (Jongman 1995; Foster 2001; McKinney, et al. 2002), bioregionalism (Miller 1996; Brunckhorst 2000), landscape ecology (Liu and

Taylor 2002; Bissonette and Storch 2003), and political science (Young 1994; Ward 1998) calls not only for the consideration of larger spatial and temporal scales, but also for the reconciliation of the natural and social sciences.

The primary driver for the convergence of the literature cited above is the recognition that existing management approaches and institutional structures have largely failed to solve problems that transcend administrative and disciplinary boundaries (Brunnée and Toope 1995; Quinn 2002). There is growing acknowledgment that environmental resources 'must be managed as parts of extensive socio-ecological systems in which environmental and human social processes are intricately related and characterized by flux and uncertainty' (Sick 2002, 1). Although the need for better integration is clearly established in the environmental management literature (Margerum and Born 1995; Hooper, et al. 1999), the challenges associated with translating the ideas on the ground have been tremendous.

The rising trend towards management of natural resources at broader landscape and regional scales has accelerated the international interest in transboundary natural resource management (Parris 2004). TBNRM has been defined as 'any process of cooperation across boundaries that increases the effectiveness of attaining natural resource management or biodiversity conservation goals' (van der Linde et al. 2001, 10). TBNRM is pursued in a wide variety of contexts with an equally diverse number of goals and objectives. The integration of policy, planning, and management across borders can result in more effective environmental management than would be the case if participants were to pursue independent programs on their respective sides of the border.

TBNRM often includes the existence or creation of transboundary protected areas (TBPA), including biosphere reserves, as core elements of the process. TBPAs have a history that extends back to the early part of the twentieth century. The establishment of a framework for border park management between Poland and Czechoslovakia in 1925 (Thorsell 1990) led to Pieniny International Landscape Park in 1932 (Turnock 2002), the creation of Albert National Park, which spans the colonial states of Ruanda-Urundi and the Congo, in 1925 (van der Linde et al. 2001), and the world's first International Peace Park, between Waterton Lakes National Park (Canada) and Glacier National Park (United States) in 1932 (Sandwith et al. 2003).

TBNRM is an integrated approach to management that often encompasses biodiversity conservation and regional development. The integration of conservation and development is generally expressed through programs labelled as 'community-based natural resource management' (CBNRM) or 'integrated conservation and development projects' (ICDPs), especially in developing countries. However, critics of CBNRM and ICDPs have suggested that the failure of such programs to adequately protect environmental resources necessitates a strengthening of centralized and top-down approaches to planning and management (Oates 1999; Terborgh 1999). In turn, advocates of more people-oriented conservation and development fear a return to the authoritarian protectionist conservation paradigm that characterized the establishment of protected areas in the era of colonial expansion (Jones and Conguica 2001; Katerere, Hill, and Moyo 2001; Brosius and Russell 2003; Wilshusen et al. 2002; Fall

2003; Wolmer 2003). The future success of TBNRM initiatives will entail the development of theory and practice to integrate meaningful local-scale (bottom-up) participation with regional-scale (top-down) planning and management (Lanfer et al. 2003). Institutional arrangements that adequately address stakeholder value conflicts will be the cornerstone of successful efforts (Agrawal and Gibson 1999; Gibson, et al. 2000).

The international expansion of TBNRM initiatives has been rapid and significant, especially with respect to TBPAs. In 1997 a comprehensive survey reported 136 complexes of internationally adjoining protected areas that in turn contained 488 individual protected areas in 98 different countries (Zbicz and Green 1997). In 2001 this number had grown to 169 complexes involving 650 individual protected areas and 113 different countries (Zbicz 2001). The level of cooperation and integration varies greatly across the adjoining protected-area complexes. Zbicz (1999) devised six categories to characterize the level of interaction between adjoining protected areas, from no cooperation through to full cooperation and integration. Eighteen per cent of the 136 complexes reported no cooperation, 39 per cent cooperated only at the level of information exchange, 33.5 per cent cooperated at a higher level (these included the categories of consultation, collaboration, and coordination of planning), and 8 per cent reported fully integrated management. There is currently no comprehensive catalogue of TBNRM initiatives that do not include a TBPA. However, as an indicator, 197 of the 302 (65 per cent) multilateral environmental agreements executed worldwide between 1972 and 2001 are focused on regional and transboundary integration (UNEP 2001).

CASE STUDIES

The Crown of the Continent, North America

The 'Crown of the Continent' (Crown) has come to refer to a region of approximately 42,000 km^2 that straddles the borders of Montana, British Columbia, and Alberta in the central portion of the North American Rocky Mountains. The Crown is split north to south by the continental divide and forms the headwaters of 19 major rivers that comprise part of three continental drainages. Internationally recognized for its ecological and geological uniqueness, the region constitutes one of the most intact and ecologically diverse areas on the continent (Long 2002). The ecological significance of the region is perhaps best indicated by the occurrence of eight large carnivore species and their associated prey; the Crown is the only area remaining in the lower 48 states where such a fully intact assemblage exists.[2] The valleys of the Crown serve as important corridors for wildlife movement, connecting populations and metapopulations of various species throughout the Rocky Mountain cordillera. Many small mammals, birds, reptiles, amphibians, and fish and a wide diversity of plants punctuate the ecological importance of the Crown.

Ancestors of the current First Peoples occupied lands of the Crown for over 10,000 years. The Crown remains a sacred landscape for the K'tunaxa bands west of the divide and tribes of the Nitsitappi (the Blackfoot Nation) on the east side. Today the

Crown is characterized by rapid urban and rural-residential expansion and by economic growth associated with the high amenity values of the region. In addition, there is substantial development in the petroleum, forestry, and agricultural sectors (Prato and Fagre 2007).

The transboundary Crown of the Continent region is characterized by complex jurisdictional fragmentation. The lands of the Crown fall under the jurisdiction of more than 20 agencies, including administrations from two provinces, one state, and two federal governments, along with First Nations, municipal governments, and a large number of private interests. Historically, natural resource decision making in the region has been fragmented and incremental, with each jurisdiction and resource sector operating in relative isolation. Approximately 60 per cent of the Crown is public land, and half of this is in some form of legislated protected area. The Waterton-Glacier International Peace Park forms the core of the protected area. International recognition of the region's significance also exists by way of UNESCO 'world heritage site' and 'biosphere reserve' designations in and around Waterton Lakes and Glacier National Parks.

Many international agreements and working arrangements currently exist between parties in the Crown, but none of these attempt overall regional integration of environmental management. In 2001 government representatives from over 20 agencies convened a meeting to explore ecosystem-based ways of collaborating on shared issues in the Crown. Participants included resource managers who represented federal, Aboriginal, provincial, and state agencies with significant land or resource management responsibility in the Crown. Government managers wanted a forum in which they could communicate and coordinate their activities without the presence of environmental advocacy organizations. The Miistakis Institute for the Rockies, a non-profit research organization affiliated with the University of Calgary, was invited to facilitate the meeting and act as a neutral third party. The highly successful founding workshop, hosted by the Waterton-Glacier International Peace Park, resulted in an informal commitment by all participants to move forward collaboratively on integrated regional management. The objectives of the CMP (Crown of the Continent Managers Partnership) are to (1) build awareness of common interests and issues in the Crown of the Continent region, (2) build relationships and opportunities for collaboration across mandates and borders, and (3) identify collaborative work already underway and opportunities for further cooperation. The CMP came to a consensus around five strategic issues of shared interest that are best addressed at a regional scale:

1. to address cumulative effects of human activity across the region;
2. to address increased public interest in how lands are managed and how decisions are reached;
3. to address increased recreational demands and increased visitation;
4. to collaborate in sharing data, standardizing assessment, and monitoring methodologies; and
5. to address the maintenance and sustainability of shared wildlife populations.

The CMP has made modest progress since its inception (see www.rockies.ca/cmp). The core of the CMP initiative has been an annual forum of the membership to discuss transboundary issues of common concern. Hosted by the International Peace Park, this annual forum has focused on a different theme each year (e.g., cumulative effects, fire management, and exotic species). To encourage applied research in the region, the CMP has fostered positive relationships with regional universities; it has had, for example, both a Canadian and an American academic representative on its steering committee. The CMP remains a government agency initiative with no direct participation from regional or local environmental non-governmental organizations (NGOs). The strategic issue of addressing shared wildlife populations has not been explicitly addressed by the CMP. However, a recent graduate research project on transboundary wildlife management (Grant 2005) and the consideration given wildlife in the regional cumulative effects assessment project (see below) illustrate the CMP's commitment to the issue.

A regional cumulative effects assessment project was initiated as one of the first concrete projects of the CMP (Quinn et al. 2005). Understanding the cumulative effects of segregated and incremental decision making was seen as a critical foundation for promoting and achieving greater integration. The realization of this ambitious project has been thwarted by a host of factors that exemplify the difficulties of working in complex, multi-jurisdictional environments. These factors include difficulties with data availability, quality, and compatibility; differing perceptions and expectations concerning the project's goals and objectives; changes in institutional structures and in personnel; lack of financial resources; lack of full participation by all partners; difficulties in developing a framework for a regional-scale assessment; absence of a formal multilateral agreement; and a lack of political will. The challenges associated with the CMP regional cumulative effects project have exposed the barriers to integration inherent in current institutional structures. If the CMP is to function as more than a forum for inter-agency communication, all participating agencies must demonstrate a greater level of commitment to creatively overcoming the barriers to integration. In order for this to happen, agencies will need to be convinced that the effort-to-reward ratio merits their full participation. They must also be given greater political support and a clear mandate for their activities. In addition, the CMP may want to consider creating a forum for communication and collaboration with NGOs such as those affiliated with the Yellowstone to Yukon (Y2Y) initiative.

Carpathian Mountains, Europe

The Carpathian Mountains form a 1,500-km arc encompassing approximately 210,000 km^2 of east-central Europe, including parts of the Czech Republic, Slovakia, Poland, Hungary, Ukraine, and Romania, as well as small portions of Austria and Serbia and Montenegro.[3] The region, known as the 'Green Backbone of Eastern and Central Europe,' provides an ecological bridge for the dispersal of biota between the northern and southern forests of the continent (Nowicki 1998; CEI 2001). The Carpathian Mountains are crucial in providing water, biological diversity, energy, minerals, forest, and agricultural products to the region. Plant diversity is of international significance, with approximately one-third of all European species (i.e., 3,800 vascular plant species) and some 200 endemic species occurring in the Carpathians. The region is approximately 50 per cent forested with some of the largest remaining tracts of European forests (especially beech) outside of Russia. The Carpathians support Europe's largest populations of large carnivores, including brown bear (*Ursus arctos*, ~8,000), grey wolf (~4,000), and Eurasian lynx (*Lynx lynx*, ~3,000), and therefore contain the most robust source populations for dispersal and restoration in Europe (Samec 2002). In addition, the region supports a suite of large mammals, including chamois (*Rupicapra rupicapra*) and recovering populations of once-extirpated European bison (*Bison bonasus*).

The Carpathians have a rich and diverse human cultural history extending back to the Pleistocene (Nelson 2004). The dynamic complexity of human activity and ethnic diversity has strongly influenced the character of the landscape. For example, high-elevation grasslands and meadows are a consequence of over 500 years of forest clearing for sheep grazing. Today, the age-old traditions of highland peoples continue to define the essence of the region. The population of the Carpathians is estimated to be between 16 and 18 million people.

Approximately 16 per cent of the Carpathian ecoregion is under some form of protection, mainly through national parks and biosphere reserves (Voloscuk 1999). Land ownership in the Carpathians is complex and has been evolving with European reunification. Lands that were isolated and depopulated during the Communist era are being reclaimed and re-inhabited. The privatization of former state-owned lands and the alterations in rural economies are having significant effects on the Carpathian landscape. The profound political, social, and economic changes that characterize the region offer both significant threats to and opportunities for sustainability. Regionally, there is 'a mood for dialogue to enhance public awareness of threats to biodiversity and embrace initiatives for more sustainable development' (Turnock 2002, 47). Furthermore, there is growing capacity in the European Union for environmental protection and sustainability through a broad array of programs, initiatives, and funding mechanisms (European Commission 2002).

The Carpathian Ecoregion Initiative (CEI, now known as CERI)[4] was initiated in 1999 as part of the World Wide Fund for Nature's (WWF) ecoregional conservation approach, the Global 200 (Olson et al. 2000). The Carpathian region was identified as the best representation of European-Mediterranean montane mixed forests and was classified as an endangered ecoregion. The WWF was already active in the area through the Danube-Carpathian Programme (CDP) and expanded its activity to include a comprehensive ecoregional conservation program for the Carpathians. Following a compilation of preliminary background information, a workshop was convened in May 1999 with participants from more than 50 government and non-government agencies. Recognizing that an integrated, international effort would be required to protect the natural and cultural capital of the Carpathians, participants agreed to move forward with an international transboundary effort that followed the WWF methodology for ecoregion-based conservation (Dinerstein et al. 2000).

An integrated framework was established to facilitate the transboundary process. The WWF Danube-Carpathian Programme office in Vienna coordinated the initiative, forming a steering group and a coordination group with the participation of representatives (country coordinators) from each participating nation. International working groups were established to oversee the biodiversity and socio-economic assessments. In addition, working groups that focused on sustainable development and communications were established. A significant element of the process was the development of a comprehensive geographic information system (GIS) approach to data collection, management, and analysis (Nelson 2004). There was considerable adaptation, coordination, and communication among the working groups. The first three phases of the ecoregional approach (reconnaissance, biodiversity assessment, and socio-economic assessment) were completed by the end of 2001 (CEI 2001) and resulted in the identification of 30 priority areas for conservation.

In April 2001 a political summit co-chaired by the president of Romania and HRH Prince Phillip (on behalf of the WWF) was held in Bucharest, Romania, to address environmental and sustainable development issues for the Carpathians and Danube. The meeting was attended by over 900 high-level participants representing the interests of the region. A refined vision for CEI was circulated among the participants, and the summit resulted in the Declaration on Environment and Sustainable Development in the Carpathian and Danube Region, which was signed by 14 heads of state.[5] The declaration highlighted the agreement to develop collaborative and integrated approaches to transboundary, ecosystem-based management for the region. The high level of political support provided the necessary encouragement and mandate for regional managers and NGOs to work together in developing and implementing a strategy.

The declaration provided the foundation for the Framework Convention on the Protection and Sustainable Development of the Carpathians (hereafter, the Convention), which was signed by representatives of all seven of the Carpathian countries in May 2003.[6] This landmark document commits the parties to 'pursue a comprehensive policy and cooperate for the protection and sustainable development of the Carpathians with a view to *inter alia* improving the quality of life, strengthening local economies and communities, and conservation of natural values and cultural heritage.' The Convention promotes achieving its aims through the application of the precautionary principle, the 'polluter pays' principle, public participation and stakeholder involvement, transboundary cooperation, integrated planning and management of land and water resources, a programmatic approach, and an ecosystem approach.

Issues related to transboundary wildlife management are covered in article 4 of the Convention. Specific reference is made to the maintenance of connectivity in the protection of endangered species, endemic species, and large carnivores. Mechanisms for integrated wildlife management include programs for the management of exotic species, promotion of collaborative research, monitoring and data sharing, completion of a Carpathian network of protected areas, and the integration of conservation and sustainable-use measures. The CERI has partnered with the Carpathian Large Carnivore

Project (CLCP),[7] a 10-year initiative founded in Romania in 1993 by the Munich Wildlife Society and the Romanian State Forest Administration. The goal of the CLCP is to implement community-based conservation in the project area through an integrated management approach. The CLCP defines its approach to integrated wildlife management as 'one that considers the inter-dependency of four fields of activity: research, management, rural development and public education. The four components form a holistic design, are closely linked and support each other, e.g., through financing and creating transparency and acceptance. The project is designed to achieve long term financial sustainability for the community' (Promberger 2001, 4).

The CLCP works with local communities to better understand and manage the conflict between people and large carnivores. Increased revenue is being generated in and by the community through ecotourism development. Tourism has been increasing by 50–120 per cent annually, and a large carnivore education centre is planned for the community of Zarnesti, Romania, near Piatra Craiului National Park. Overall, the program is an exemplary international model of an integrated approach to wildlife and community sustainability research, management, development, and education. The experience of the CLCP is now being used in the development of the Carpathian Action Plan for the Large Carnivores, a pragmatic approach to maintaining carnivore populations under the auspices of the Convention.

The implementation phase of the Carpathian Ecoregion Initiative under the mandate of the Convention was initiated at the Brasov Multistakeholder Workshop held in Romania in November 2003. In August 2004, following some difficulties between the CEI and the WWF, the organization reconstituted itself as a politically independent, non-profit, non-governmental organization, registered in the Slovak Republic with headquarters at the Daphne Institute for Applied Ecology in Bratislava. The Secretariat of the CERI (formerly CEI) will remain at the WWF-Danube-Carpathian Programme in Vienna on an interim basis. The CERI is moving forward with the development and implementation of a Multistakeholder Ecoregion Action Plan via a Global Environmental Facility project for the Carpathians with the United Nations Development Program, the United Nations Environment Program, and the World Bank. The CERI will function as an independent, non-government network to broker conservation with local, regional, and international agencies and interests.

LESSONS LEARNED AND FUTURE DIRECTIONS

Transboundary wildlife management has benefited from new technologies such as satellite collars for monitoring wildlife movement, improved resolution of remotely sensed imagery, and computers' increased capacity and accessibility for spatial analysis. The transboundary movement of wildlife is a high-profile integrator that can act as a catalyst for collaboration across borders. Data and subsequent analyses from wildlife movement studies have reinforced the need for integrated approaches to management that encompass larger spatial scales and longer time frames. The development of common research and monitoring protocols and the subsequent sharing of data may create logistical challenges, but these are outweighed by the synergistic benefits that can accrue to participants. Transboundary protected areas can provide an essential foundation for ecoregional approaches to wildlife management and regional development.

Meaningful integrated management of wildlife across boundaries will require the removal of structural obstacles inherent in the policy, legislation, and institutional frameworks that currently characterize the fragmented jurisdictional systems of management (Kennett 2002). No amount of inter-agency cooperation can fully overcome the effects of decision-making processes that are rooted in narrowly focused mandates. To rise above institutional barriers successfully will require a high level of political support manifested in multilateral, international agreements and working arrangements. Furthermore, the active engagement and leadership of local, regional, and international NGOs is essential to integrated approaches to wildlife management. These organizations can provide a facilitation and advocacy role that is not generally possible within the confines of government agency structures and four-year political cycles. Ultimately, integration across administrative boundaries is about human inter-relationships (Zbicz 2001). The influence of individual personalities, strong leadership, and dedication in overcoming barriers cannot be overstated. A commitment to communication and face-to-face meetings is the cornerstone of effective integration across boundaries. Finally, large-scale integrated efforts in the management of transboundary wildlife will require both the top-down support described above and a bottom-up approach that meaningfully includes the values of the local communities within the overall region of interest.

NOTES

1. Long-range movements by radio-collared wolves were seminal in the formation of the Yellowstone to Yukon (Y2Y) Conservation Initiative, a non-governmental vision, strategy, coalition, and organization focused on continental-scale conservation and ecological connectivity in the Rocky Mountain ecoregions of North America (Tabor 1996; see http://www.y2y.net).
2. Mountain caribou (*Rangifer tarandus caribou*) and bison (*Bison bison bison*) were extirpated from the system early in the last century.
3. The percentage of the Carpathians in each country are as follows: Austria, <1%; Czech Republic, 3%; Serbia and Montenegro, <1%; Slovakia, 17%; Poland, 10%; Hungary, 4%; Ukraine, 11%; and Romania, 55%.
4. The program was initially known as the Carpathian Ecoregion Based Conservation Program. The name was subsequently shortened to the Carpathian Ecoregion Initiative, and the program went by the acronym CEI. Most recently, the program is identified by the acronym CERI. For more information, see http://www.carpathians.org/.
5. For the full text of the declaration, see http://www.carpathians.org/sum_info1.htm.
6. For the full text of the framework convention, see http://www.ceeweb.org/a6carpathian/docs/carpathian.conv.pdf.
7. The CLCP is part of the pan-European Large Carnivore Initiative for Europe, a dynamic network of representatives from governments, international and national NGOs, scientists,

and other experts working across Europe to promote the coexistence of brown bears, lynxes, wolves, and wolverines with human societies. See http://www.lcie.org/.

REFERENCES

Agrawal, A., and C. Gibson. 1999. Enchantment and disenchantment: The role of community in natural resource conservation. *World Development* 27 (4): 629–49.

Bennett, A.F. 2003. *Linkages in the landscape: The role of corridors and connectivity in wildlife conservation.* Gland, Switzerland: IUCN.

Berkes, F., and C. Folke. 1998. *Linking social and ecological systems: Management practices and social mechanisms for building resilience.* Cambridge, UK: Cambridge University Press.

Bissonette, J.A., and I. Storch. 2003. *Landscape ecology and resource management: Linking theory with practice.* Washington, DC: Island Press.

Boyd, D.K., and D.H. Pletscher. 1999. Characteristics of dispersal in colonizing wolf population in the central Rocky Mountains. *Journal of Wildlife Management* 63:1094–108.

Brosius, J.P., and D. Russell. 2003. Conservation from above: An anthropological perspective on transboundary protected areas and ecoregional planning. *Journal of Sustainable Forestry* 17 (1/2): 35–58.

Brunckhorst, D.J. 2000. *Bioregional planning: Resource management beyond the new millennium.* Amsterdam: Harwood Academic.

Brunnée, J., and S.J. Toope. 1995. Environmental security and freshwater resources: A case for international ecosystem law. In G. Handl (ed.), *Yearbook of international environmental law*, 41–76. Oxford: Clarendon Press.

CEI (Carpathian Ecoregion Initiative). 2001. *The status of the Carpathians.* Vienna: WWF International, Danube-Carpathian Programme.

Chester, C.C. 2006. *Conservation across borders: Biodiversity in an interdependent world.* Washington, DC: Island Press.

Dinerstein, E., G. Powell, D. Olson, E. Wikramanayake, R. Abell, C. Loucks, E. Underwood, T. Allnutt, W. Wettengel, T. Ricketts, H. Strand, S. O'Connor, and N. Burgess. 2000. A workbook for conducting biological assessments and developing biodiversity visions for ecoregion-based conservation. Part I: Terrestrial ecoregions. Washington, DC: WWF-US, Conservation Science Program.

European Commission. 2002. *Choices for a greener future: The European Union and the environment.* Luxembourg: Office for Official Publications of the European Communities.

Fall, J. 2003. Planning protected areas across boundaries: New paradigms and old ghosts. *Journal of Sustainable Forestry* 17:81–102.

Forman, R.T.T. 1995. *Land mosaics: The ecology of landscapes and regions.* Cambridge, UK: Cambridge University Press.

Forman, R.T.T., and P.N. Moore. 1992. Theoretical foundations for understanding boundaries in landscape mosaics. In A.J. Hansen and F. di Castri (eds), *Landscape boundaries: Consequences for biotic diversity and ecological flows*, 236–59. New York: Springer-Verlag.

Foster, K.A. 2001. *Regionalism on purpose.* Cambridge, MA: Lincoln Institute of Land Policy.

Gibson, C., M.A. McKean, and E. Ostrom. 2000. *People and forests: Communities, institutions, and governance.* Cambridge, MA: MIT Press.

Glick, D.A., and T.W. Clark. 1998. Overcoming boundaries: The Greater Yellowstone Ecosystem. In R.L. Knight and P.B. Landres (eds), *Stewardship across boundaries*, 237–56. Washington, DC: Island Press.

Grant, J. 2005. *Driving forces and barriers to transboundary wildlife management: The Crown of the Continent ecosystem experience*. M.Sc. thesis, Interdisciplinary Graduate Program, University of Calgary.

Griffin, J., D. Cumming, S. Metcalfe, M. t'Sas-Rolfes, J. Singh, E. Chonguiça, M. Rowan, and J. Oglethorpe. 1999. *Study on the development of transboundary natural resource management areas in southern Africa*. Washington, DC: Biodiversity Support Program.

Groves, C.R. 2003. *Drafting a conservation blueprint: A practitioner's guide to planning for biodiversity*. Washington, DC: Island Press.

Grumbine, R.E. 1994. What is ecosystem management? *Conservation Biology* 8:27–38.

Hamilton, L.S. 1996. Transborder protected area co-operation. In J. Cerovský (ed.), *Biodiversity conservation in transboundary protected areas in Europe*, 9–18. Prague, Czech Republic: Ecopoint.

Hobbs, J.C.A. 1981. The environmental impact of veterinary cordon fences. *African Wildlife* 35 (6): 16–21.

Hooper, B.P., G.T. McDonald, and B. Mitchell. 1999. Facilitating integrated resource and environmental management: Australian and Canadian perspectives. *Journal of Environmental Planning and Management* 42 (5): 747–66.

Inman, R.M., R.R. Wigglesworth, K.H. Inman, M.K. Schwartz, B.L. Brock, and J.D. Rieck. 2004. Wolverine makes extensive movements in the Greater Yellowstone Ecosystem. *Northwest Science* 78 (3): 261–6.

Jennings, J.J., and Scott, J.M. 1993. Building a macroscope: How well do places managed for biodiversity match reality? *Renewable Resources Journal* 11 (2):16–20.

Jones, B.T.B, and E. Chonguiça. 2001. Review and analysis of specific transboundary natural resource management initiatives in the Southern African Region. Paper No.2, IUCN-ROSA Series on Transboundary Natural Resource Management. Harare, Zimbabwe: IUCN.

Jongman, R.H.G. 1995. Nature conservation planning in Europe: Developing ecological networks. *Landscape and Urban Planning* 32:169–83.

Katerere, Y., R. Hill, and S. Moyo. 2001. *A critique of transboundary natural resource management in Southern Africa*. Paper No.1, IUCN-ROSA Series on Transboundary Natural Resources Management. Harare, Zimbabwe: IUCN.

Kennett, S.A. 2002. *Integrated resource management in Alberta: Past, present and benchmarks for the future*. CIRL Occasional Paper No. 11. Calgary: Canadian Institute for Resources Law.

Landres, P.B., R.L. Knight, S.T.A. Pickett, and M.L. Cadenasso. 1998. Ecological effects of administrative boundaries. In R.L. Knight and P.B. Landres (eds), *Stewardship across boundaries*, 39–64. Washington, DC: Island Press.

Lanfer, A.F., M.J. Stern, C. Margoluis, and U.M. Goodale. 2003. A synthesis of the March 2001 Conference on the Viability of Transboundary Protected Areas at the Yale School of Forestry and Environmental Studies. *Journal of Sustainable Forestry* 17 (1/2): 229–42.

Liu, J., and W.W. Taylor. 2002. *Integrating landscape ecology into natural resource management*. Cambridge, UK: Cambridge University Press.

Long, B. 2002. *Crown of the Continent: Profile of a treasured landscape*. Kalispell, MT: Crown of the Continent Ecosystem Education Consortium.

McKinney, M., C. Fitch, and W. Harmon. 2002. Regionalism in the west: An inventory and assessment. *Public Land and Resources Law Review* 101:101–20.

Madonna, K.J. 1995.The wolf in North America: Defining international ecosystems vs. defining international boundaries. *Journal of Land Use and Environmental Law* 10 (2): 1–38.

Margerum, R.D., and S.M. Born. 1995. Integrated environmental management: Moving from theory to practice. *Journal of Environmental Planning and Management* 38 (3): 371–91.

Merrill, S.B., and D.L. Mech. 2000. Details of extensive movements by Minnesota wolves (*Canis lupus*). *American Midland Naturalist* 144:428–33.

Miller, K. 1996. *Balancing the scales: Guidelines for increasing biodiversity's chances through bioregional management.* Washington, DC: World Resources Institute.

Musiani, M., and P.C. Paquet. 2004. The practices of wolf persecution, protection and restoration in Canada and the United States. *BioScience* 54 (1): 50–60.

Nelson, J.G. 2004. *The Carpathians: Assessing an ecoregional planning initiative.* Waterloo, ON: Environments Publication, Faculty of Environmental Studies, University of Waterloo.

Noss, R.F., and A. Cooperrider. 1994. *Saving nature's legacy: Protecting and restoring biodiversity.* Washington, DC: Island Press.

Nowicki, P. (ed.). 1998. *The green backbone of central and eastern Europe.* Tilburg: European Centre for Nature Conservation.

Oates, J.F. 1999. *Myth and reality in the rainforest: How conservation strategies are failing in West Africa.* Berkeley, CA: University of California Press.

Olson, D.M., E. Dinerstein, R. Abell, T. Allnutt, C. Carpenter, L. McClenachan, J. D'Amico, P. Hurley, K. Kassem, H. Strand, M. Taye, and M. Thieme. 2000. The Global 200: A representation approach to conserving the Earth's distinctive ecoregions. Washington, DC: World Wildlife Fund–US, Conservation Science Program.

Paquet, P.C. 2005. Personal communication.

Parris, T.M. 2004. Managing transboundary environments. *Environment* 46 (1): 3–4.

Prato, T., and D. Fagre (eds). 2007. *Sustaining Rocky Mountain landscapes: Science, policy and management for the Crown of the Continent Ecosystem.* Washington, DC: RFF Press.

Promberger, C. 2001. *The integrated management approach in wildlife conservation field projects: Experiences and case study.* Romania: Carpathian Wildlife Foundation.

Quinn, M.S. 2002. Ecosystem-based management. In D. Thompson (ed.), *Tools for environmental management: A practical introduction and guide*, 370–82. Gabriola Island, BC: New Society Press.

Quinn, M., G. Greenaway, D. Duke and T. Lee. 2005. *A collaborative approach to assessing regional cumulative effects in the transboundary Crown of the Continent.* Canadian Environmental Assessment Agency, Research and Development Monograph Series. Ottawa: Canadian Environmental Assessment Agency.

Redmond, C.L. 1999. Human dimensions of ecosystem studies. *Ecosystems* 2:296–8.

Reid, W.V. 1996. Beyond protected areas: Changing perceptions of ecological management objectives. In R.C. Szaro and D.W. Johnston (eds). *Biodiversity in managed landscapes: Theory and practice*, 442–53. New York: Oxford University Press.

Sale, K. 1985. *Dwellers in the land: The bioregional vision.* San Francisco: Sierra Club.

Samec, M.E. 2002. Existing instruments and programmes and a Carpathian sector analysis. Background paper intended for the negotiation process towards a Carpathian Framework Convention. Vienna: United Nations Environmental Program.

Sanderson, E.W., K.H. Redford, A. Vedder, S.E. Ward, and P.B. Coppolillo. 2002. A conceptual model for conservation planning based on landscape species requirements. *Landscape and Urban Planning* 58:41–56.

Sandwith, T., C. Shine, L. Hamilton, and D. Sheppard. 2003. *Transboundary protected areas for peace and co-operation.* Gland, Switzerland: IUCN.

Scott, J.M., E.A. Norse, H. Arita, A. Dobson, J.A. Estes, M. Foster, B. Gilbert, D.B. Jensen, R.L. Knight, D. Mattson, and M.E. Soulé. 1999. The issue of scale in selecting and designing biological reserves. In M.E. Soulé and J. Terborgh (eds), *Continental conservation: Scientific foundations of regional reserve networks*, 19–37. Washington, DC: Island Press.

Sick, D. 2002. *Managing environmental processes across boundaries: A review of the literature on institutions and resource management.* Report prepared for the Minga Program Initiative. Ottawa: International Development Research Council.

Soulé, M.E., and J. Terborgh. 1999. *Continental conservation: Scientific foundations of regional reserve networks.* Washington, DC: Island Press.

Tabor, G.M. 1996. *Yellowstone-to-Yukon: Canadian conservation efforts and continental landscape/biodiversity strategy.* Boston: Henry P. Kendall Foundation.

Terborgh, J. 1999. *Requiem for nature.* Washington, DC: Island Press.

Thorsell, J. 1990. Through hot and cold wars, parks endure. *Natural History* 99 (6): 56–9.

Turnock, D. 2002. Ecoregion-based conservation in the Carpathians and the land-use implications. *Land Use Policy* 19:47–63.

UNEP (United Nations Environmental Program). 2001. Report of the Executive Director, Global Ministerial Environment Forum, International Environmental Governance. Seventh special session, Cartagena, Colombia, 13–15 February 2002, UNEP/GCSS, VII/2, p. 15.

van der Linde, H., J. Oglethrope, T. Sandwith, D. Snelson, and Y. Tessema. 2001. *Beyond boundaries: Transboundary natural resource management in sub-Saharan Africa.* Washington, DC: Biodiversity Support Program.

Vangen, K.M., J. Persson, A. Landa, R. Andersen, and P. Segerstrom. 2001. Characteristics of dispersal in wolverines. *Canadian Journal of Zoology* 79:1641–9.

Voloscuk, I. (ed.). 1999. *The national parks and biosphere reserves in Carpathians.* Tatranska Lomnica, Slovak Republic: Association of the Carpathian National Parks and Biosphere Reserves.

Ward, V. 1998. Sovereignty and ecosystem management: Clash of concepts and boundaries? In K. Litken (ed.), *Environment and Sovereignty*, 79–108. Cambridge, MA: MIT Press.

Westing, A.H. 1993. Biodiversity and the challenge of national borders. *Environmental Conservation* 20 (1): 5–6.

Westing, A.H. 1998. Establishment and management of transfrontier reserves for conflict prevention and confidence building. *Environmental Conservation* 25 (2): 91–4.

Wilshusen, P.R., S.R. Brechin, C.L. Fortwangler, and P.C. West. 2002. Reinventing a square wheel: Critique of a resurgent 'protection paradigm' in international biodiversity conservation. *Society and Natural Resources* 15:17–40.

Wolmer, W. 2003. Transboundary conservation: The politics of ecological integrity in the Great Limpopo Transfrontier Park. *Journal of Southern African Studies* 29 (1): 261–78.

WWF (World Wide Fund for Nature). 2001. Carpathian Ecoregion Initiative: Bringing people and nature together in the heart of Europe. Press invitation. Vienna: WWF International, Danube-Carpathian Programme Office. URL: http://www.carpathians.org/docs/Press%20briefing%20-%20CEI.doc, accessed 16 November 2004.

Young, O.R. 1994. The problem of scale in human/environment relationships. *Journal of Theoretical Politics* 6 (4): 429–47.

Zbicz, D.C. 1999. *Transboundary cooperation in conservation: A global survey of factors influencing cooperation between internationally adjoining protected areas.* Ph.D. dissertation, Duke University, Durham, NC.

——— 2001. Crossing boundaries in park management. In D. Harmon (ed.), *Proceedings of the 11th Conference on Research and Resource Management in Parks and on Public Lands*, 197–203. Hancock, MI: George Wright Society.

Zbicz, D.C., and M.J.B. Green. 1997. Status of the world's transfrontier protected areas. *Parks* 7 (3): 5–10.

Conclusion: Reflections and Prospects for IREM

D. Scott Slocombe and Kevin S. Hanna

This book was not intended to provide a definitive description, definition, or prescription for integrated resource and environmental management (IREM)—if that were even possible. We do hope, however, that it has provided some historical and conceptual context for current efforts toward integration and some examples of best practice in a range of areas of application. In that spirit we want now to highlight and comment on some critical themes, challenges, and opportunities for moving integrated resource and environmental management forward. No doubt others might identify different lessons and points to emphasize.

Integration has a long history, under that name and many others, in resource management, watershed management, and coastal zone management, among others. Good questions are, what has changed, what hasn't, and why the reinvention? Clearly, a big part of what has changed over the last 40 years is the complexity of the resource and environmental management (REM) context and our understanding of it. We now understand much more fully, and respond more appropriately, to multiple demands, multiple scientific and other perspectives, multiple stakeholders, and the inherent complexity and uncertainty in resource and environmental management. The chapters in this book reflect this fact and illustrate its implications in numerous specific ways, some of which we highlight below. As the REM context changes over time and from place to place and as our understanding of it changes, so must integration theory and practice change. As long as we don't ignore history, as long as we do learn from it and don't repeat too much of it, a bit of repackaging and reinvention may not be a bad thing. Such an approach is almost certainly necessary as circumstances change. And we now have more, and multiple, methods for responding to the challenges of integrated resource and environmental management.

Given the uniqueness and changeability of REM contexts in space and time, local characteristics and dynamics become very important in a range of ways: as a focus for different perspectives and developing understandings of problems; as a locale for both formal and informal economies and politics; and as a starting point for participation and collaboration.

Informal and formal, personal and organizational collaboration and participation are the most ubiquitous themes of these papers. Their general ubiquity underscores their importance. The diversity of specifics—the range of activities that might fall within and between the two terms—however, underscores our need to consider sociopolitical contexts, management goals, and processes in order to tailor participation and collaboration to each problem, process, or place. Socio-economic and political structures and dynamics are also highly relevant to participation and collaboration, since this is where participation and collaboration are significant issues, although we might also consider how to effectively foster participation of the distant in the local and of the local in the distant.

Linked to issues of formal and informal economies and politics, and of local versus distant actors, are issues of power and equity. Several chapters in this volume illustrate the challenges of recognizing, working with, and supporting the resource and environmental activities of more or less disenfranchised groups. These remain challenges for resource and environmental management broadly, and specific approaches and awarenesses, beyond integration, are needed to address them.

There can be little doubt that, ideally, a wide range of information is needed for IREM—at the very minimum, natural and social sciences as well as historical, local, and experiential knowledge and understanding. The acquisition of such information has strong local dimensions, as it can be gained through local participation and listening, through local collaboration in data collection and analysis, and in the joint production of sound, credible, and acceptable interpretations and assessments. Information technology (IT) plays a significant role: remote sensing assists in data collection; and databases and the geographic information system (GIS) greatly assist with organizing and interpreting the wide range and large volume of information expected in modern IREM. GIS, mapping, and data analysis software facilitates full use of the information and production of highly effective graphical products for discussion, consultation, and advocacy. Websites and email can assist in communication within an IREM process and with larger-scale communication efforts. But the full use of most IT tools requires considerable expertise and funding, and these tools are no substitute for first-hand experience and knowledge of a region or for face-to-face meetings, consultation, research, and analysis. IT is a great complement to, but not a substitute for, more traditional approaches. Specific IREM exercises should give careful consideration to IT needs, tools, and roles within the context of their own goals, processes, and needs. In addition, there are deeper questions related to the integration of types and forms of information (e.g., scientific, local, Aboriginal) that are unlikely to have technological solutions.

Information has not only become more abundant in recent decades, but has also become both more long-lasting and more ephemeral. This is because, in its sheer volume and increasingly electronic forms, it is both easier to store and easier to lose. And it is hard to say in advance what information will be kept, what will be lost and then found much later, and what will be lost altogether. In such an environment, metadata are more important than ever, and this probably applies not only to quantitative or geospatial data, but also to qualitative, experiential knowledge, from traditional knowledge to knowledge gained from experience with IREM itself.

Information management and retention are related to learning. This is a relatively new topic for IREM, but an increasingly important one. With IREM, learning can take place at many levels—individual, organizational, social, or cultural—and covers a range of areas—from the biophysical context to how to facilitate effective and efficient planning, management, and assessment processes. Local and individual scales are again critical. Learning may start at the local scale and involve reflective individuals engaged in active learning. It can also benefit from opportunities to experiment, to try things out, on a small scale, as well as from higher-level, more institutionalized perspectives and comparisons of experience with similar problems in multiple places.

Integration implies both crossing boundaries and integrating across them. This book illustrates many kinds of boundaries to cross: disciplines; management units; diverse management agencies; multiple scales of biophysical, political, and administrative structures and processes; and individual perspectives and understandings. Individuals often play key roles, providing leadership to bridge scales and boundaries—as do concrete projects with cross-scalar connections and implications. It is also important that those involved maintain the necessary flexibility to respond to the change that results from crossing boundaries.

The preceding chapters offer many examples of ways to foster integration across boundaries. These include engaging in the collaborative development of knowledge of a problem or region, pursuing complex synthetic goals such as livelihoods and sustainability, and building new mechanisms of communication and management around a complex problem. None of these undertakings are simple or are likely to proceed without some tension and conflict. Thus, in addition to the need for learning, there are clear needs for improvements in communication and conflict resolution. A weakness in these areas has characterized resource and environmental theory and practice for a long time. While there have been advances in communication and conflict resolution, much remains to be done.

Beyond this, the issue of whether or not there is a need for new institutions and governance has been debated repeatedly in integrated resource and environmental approaches of many sorts. The integration of fragmented agencies and programs has been a core rationale for IREM. There can be little doubt of the need for changes in many existing institutions to facilitate IREM, but is it easier to try to reform these institutions or to start over? Of course, we rarely, if ever, start over by completely getting rid of the old; the problem is that in practice we usually retain the old and create something new as well. This may be one of the major reasons for the reinvention and recycling of approaches and ideas in REM. Another may be that we have never monitored IREM and its outcomes well enough, or systematically tracked our experience with IREM. Even the results of environmental impact assessment processes and decisions are poorly monitored in most places.

The literature on IREM participation and collaboration and many other REM approaches is rapidly expanding, and the number of practical examples is increasing. The key dimensions of IREM are being steadily explored in theory and in practice. The big question has always been how to put theory and practice together. A key lesson of this book is that there are many ways to put together myriad concepts and tools to work towards IREM—that is, towards the comprehensive integrated management of a problem or region with interconnected, multi-scale socio-economic and biophysical dimensions and complex goals, such as sustainability, ecosystem integrity, and well-being. At the same time, there are still newer approaches that seek to package and organize diverse approaches and experiences—for example, adaptive governance and adaptive co-management. We hope that we have contributed to the ongoing evolution of ideas and approaches, but we have no doubt that integration will remain a central concern and challenge for some time to come.

Index

active learning, 159
adaptive management, xiii, 105–6, 191–2; benefits of, xiii, 110; disbenefits of, xiii, 110; of watersheds, 97–118
adversarial dynamics, 122
adversary science, xiv, 182, 190
adverse conditioning, 157
agency, demands for, xiii, 57–8, 59
Agenda 21, 23–4
Alberta, xiv, 3, 236, 240
Albert National Park, 239
Al-Hawamdeh, S., 223
Alire, G., 204
Alston, R.M., 207
André, P., 77
Argyris, C., 224
Arizona, 214
Armour, A., 11, 12
Ashcroft, C., 191
Aune, K., 145

Ballard, H., xiv, 124, 164–80
Barnett, J., 53–4
Bawden, D., 221, 222, 223
Beanlands, G., 78
Bedell, J., 213
Behan, R.W., 3
benefit, xiii, 3; economic, 2, 5; social, 2
Bennett, A.F., 238
Berkes, F., 145, 231, 238
best practice sustainability, 92
best use, 25
Bhaskar, R., 57
biophysical issues, 3, 4, 6, 26, 73–96, 123, 255; and decision-making, 141–2

biosphere reserve, 241
Bird, B., 216
Bjorklund, J., 100
Blackfoot Challenge, 153, 154
Blackfoot Valley, 137–63; and Landowner Advisory Group, 155; and Waste Management and Sanitation Work Group, 155; and Wildlife Committee, 154–5
Bolland, J.M., 139
Born, S.M., 1, 4, 13, 122, 238, 239
boundaries, national, 236–51
Boundary Waters Treaty, 220
Bouthillier, F., 221, 222, 223
Bowes, M.D., 2
Boyle, J., 167
Boynton, W.R., 7, 38
Bradbury, H., 173, 174
Bradford, N., 60
Braman, S., 229–31
Brasov Multistakeholder Workshop, 245
Bregha, F., 58
Brekke, J.S., 122
Brick, P.D., 153, 154
British Columbia, xiii, 3, 81, 119–36, 240
British Columbia Ministry of Environment, 127
British North America Act, 59
Brock, W.A., 114
Brown, B.A., 164
Brundtland Commission, 23, 58
Brunner, R.D., 12, 137, 138, 139
Bucharest (Romania), 244
Bukowitz, W.R., 223
Bureau of Indian Affairs (BIA), 236

Bureau of Land Management (BLM), 236
Burns, T., 58
Burrard Inlet Environmental Action Plan, 127

Caldwell, L.K., 3, 4, 58
Cancer Incidence Study, 192–99; technical subcommittee for (TSC), 194–9
cancer risks and air emissions, xiv, 192–9
Carlson, A., 206
Carpathian Ecoregion Initiative (CEI), xiv, 238, 242–5; and Multistakeholder Ecoregion Action Plan, 245
Carpathian Large Carnivore Project (CLCP), 244–5
Carpathian Mountains, 242–5
Carpenter, S.R., 62
Carson National Forest, 204–19
catchment management, 38, 39–41, 48–53, 109
Cawley, R.M., 154
Chávez, D., 208
Chesapeake Bay, 109
Child, M. 4, 11, 12
Choo, C.W., 229
civil science, 164–80
Clapp, E.H., 208, 210
Clark, S.G., xiii, 137–63
Clark, T.W., 137–58, 238
Clary, D., 207
Clean Water Act, 25, 97, 98
clear-cutting, 26, 174
Clementsian ecology, 100
Cocksedge, W., 166
collaboration, xiv, 3–5, 10–13, 22–3, 26, 28–9, 33, 36–55, 56–8, 109, 119–36, 144, 146, 158, 252–3; among stakeholders, 181–203; cross-border, 236–51; eco-system based, 241
collaborative decision-making process, 181–203
Collins, D., 169
Columbia River, 105
communication, 109, 119–36, 145, 157, 159, 242
communication-as-integration, 45–54

community-based natural resource management (CBNRM), 239
community-level development, 92
complexity theory, 98
'comprehensive approach,' 28
conflict, human–wildlife, 137–63
conflict management, integrated approach to, 137–63
Connery, A.W., 211
Connor, R., 13, 41
consensus building, 181–203
Conservancy Districts, 27
conservation, 22, 143, 158, 244; Anglo approaches to, 216; of soil and water, 98; and wise use, 2
Conservation Authorities Act, 225, 226
Conservation Authority Information Management Strategy, 227, 231
Conservation Authorities, Ontario (see Ontario Conservation Authorities)
Conservation Ontario, 224–8
Conservation Reserve Program (CRP), 102
context/contextual, 139–63, 183–4
conventional science, 164–80
Convention on Biodiversity, 23
cooperation, xiv, 3–5, 10–13, 22–3, 26, 28–9, 33, 36–55, 56–8, 104, 119–36, 139, 159, 228; transboundary, 244
coordination, 121; and collaboration, 103, 240; and communication, 48–55; and cooperation, 28–9; and integration, 45; cross-sectoral, 99
Cornett, Z.J., 153
Cortner, H.J., 4, 24, 26
Costanza, R., 3, 7
Cottam, L.F., 211
Cronin, B., 223
Crown of the Continent, xiv, 238, 240–2
Crown of the Continent Managers Partnership (CMP), 241–2

Dale, A., xiii, 56–72
Dale, E.E., 142
Dana, S.T., 25
Danube-Carpathian Program (CDP), 243, 244

data gaps, 123, 188
decision-making processes, 181–203
Declaration on Environment and Sustainable Development in the Carpathian and Danube Region, 244
Defenders of Wildlife, 152, 156
designated operator, 207, 211, 212, 213
Deutsch, S., 206
DeWalt, E.E., 145
Diamond, J., 104
Dietz, T., 58
Dinerstein, E., 243
disbenefit, xiii, 110
Dobell, R., 59
Dood, A.R., 139, 144
Dorcey, A.H.J., 130
Dovers, S., xiii, 13, 36–55
Dryzek, J.S., 42, 125
'duelling experts,' 190
Duerr, W.A., xii, 2
Duinker, P., 78
Duke City Lumber Company, 212–15
Dunlap, L., 191
Dyrness, C.T., 171

Earth Summit, 23
Easthouse, K., 214
ecological-economic production possibilities frontier (EEPPF), 111–15
ecological economy, 43, 100
ecologically-based policies, 4, 59
ecologically sustainable development (ESD), 109–10
ecoregional-based approaches, 243, 246
ecosystem/ecosystem-based: approach, xii, 4–5, 26, 232, 244; integrity, 255; management, 202, 213, 220, 238, 241, 244
'edge,' 33
Egan, J.A., 211
Ehrmann, J.R., 190
Ellemor, H., 53–4
Emery, R.E., 57, 59, 60
endangered: ecoregion, 243; species, 202, 214
Endangered Species Act, 25, 98, 137, 139, 154, 237
energy cascading, 5
Environment Canada, 127

environmental assessment: and biophysical factors, 73–8; and community factors, 76, 78; components of, 75–6; and cultural factors, 74–8; and ecological factors, 74, 76; evolution of, 73–7; and integration, 77–80; and sustainability, 80–96
environmental degradation, 4
environmental impact assessment, xiii, 45, 73
Environmental Protection Agency (EPA), 193
Environmental Quality Incentive Program (EQIP), 102, 156
Ewart, A.W., 1, 3, 6
European Commission, 8, 243

Fall, J., 238
Farm Bill (US), 102
Fedkiw, J., 25, 26
Field, Patrick, xiv, 181–203
Finkel, M., 144
Fish and Wildlife Service (FWS), 143; 144; 147; 214
Fisheries and Oceans Canada, 127
flexible environmental assessment processes, 85
flexible management, 104, 109
flexibility in integration, 123
Folke, C., 231, 238
Forest and Rangelands Renewable Resources Planning Act, 212
forest management, xiv, 1, 26, 184, 204–19
Forest Principles, 23
forestry: and resource management, 2, 3; and water management, 24; and water resources, 2, 5
Forest Services (US), xiii, 24–6, 175, 204–19
Fortmann, L., xiv, 124, 164–80
'four-town area,' 192–9
fragmentation, 1–2, 7, 12, 45, 48, 75, 119, 125, 126, 130, 133, 138, 236, 241, 246, 254
Framework Convention on Climate Change, 23
Framework Convention on the Protection and Sustainable Development of the Carpathians, 244–5

Francis, G.R., 8
Franklin, J.F., 171
Fraser River Estuary Management Program (FREMP), xiii, 119–21, 126–36; action programs, 130; cooperation, 121; and Coordinated Project Review Process (CPRP), 131, 133; coordination, 121; and Environmental Review Committee (ERC), 131; and Estuary Management Plan (EMP), 130–1, 133; implementation structure, 129; integrated approach, 121; integration, 126–36; linked system, 127; Memorandum of Understanding (MOA), 127–8, 133, 134
Fraser River Estuary Study (FRES), 126–9, 132
Fraser River Port Authority, 127
Freeman, H., 122

Ganaraska River, 27, 28
'generalized specialists,' 8
genetically modified organisms (GMO), 98
geographic information system (GIS), xiv, 8, 113, 145–6, 152, 153, 157, 244, 253
Gibson, R.B., xiii, 10, 72–96
Glacier National Park, 239, 241
Glick, D.A., 238
Global 200, 243
Global Positioning System (GPS), 154, 236
'good management,' 5
'good science,' 182
Gottfried, R.R., 100
Goucher, N.P., xiv, 220–35
Grand River, 29–30
Grand Teton Mountains, 236
Grant, J., 242
Grayson, R.B., 105
Greater Yellowstone Ecosystem, 236
Great Falls, 146
Great Lakes, 29, 30, 220–1, 224–9, 232
Great Lakes Water Quality Agreement, 220
green accounting, 44
Green Backbone of Eastern and Central Europe, 242
Green, M.J.B., 240
Griffin, J., 237
grizzly bears, human conflict with, xiii–iv, 137–63

Gross, L.S., 211
Groves, C.R., 238
Grubb, M., 23–4
Guelph (Ontario), 27
Gunderson, L.H., 61, 99

Hanna, K.S., xiii, 8, 11, 1–20, 119–36
Hansis, R., 168
Ham, C., 122
Haraway, D., 164, 166, 167
Harvey, D., 57
Hawken, P., 97, 102
Hays, S.P., 26
Henderson, J.A., 170
Hennessey, T.M., 109
Heritage River, 30
Hessing, M., 122
Hill, M., 122
Hill, S.B., 58
Hispano residents of Vallecitos, 204–19; and environmental knowledge, 208, 212–13, 215, 216; and land-use practices, 208, 210; and poverty, 204–19; and racial positioning, 206–19
Historic Preservation Act, 25
Hobbs, J.C.A., 237
Hodge, C.L., 27, 81
Holling, C.S., 8, 56, 61, 62, 99, 105, 231
Hooper, B.P., xiii, 1, 3, 13, 33, 97–118, 239
Horlick-Jones, T., 60
Hornbeek, J., 220, 221, 229
Hovey, F.W., 143
Hovland, I., 223, 224
Howlett, M., 122
Hyder, M., 122

Illinois, 105
impact assessment, 12
implementation gap, xiii, 124
implementation imperative, 134
implementation of policy, 119–36; challenges in, 122–6, 134; framework for, 122–6, 132; multi-jurisdictional, 124, 131–2
information management, 220–35, 253; organically-developed approach to, 229; process model in, 229; systems approach to, 231

information technology (IT), 226, 253
Inglehart, R., 56
Inman, R.M., 236
integrated catchment management (ICM), 39–41
integrated conservation and development projects (ICDP), 239
integrated planning and management, 244
integrated research, xiii, 36–55, 159; and management, 36–55; and policy, 36–55
integrated resource and environmental management (IREM), 1–20, 21–35, 131, 181, 252–5; application of, 21–35; concepts of, 21–2; definitions of, 10–13, 32; dimensions of, 7–10; evolution of, 21–35; implementation of, 6–7, 21–35, 119–36; and stakeholder interests, 181–203
integrated resource management (IRM), xii, 1–20, 21–35, 58, 61, 121, 124, 164, 190, 202, 220, 226, 232; challenges and issues, xiii, 6–7; definitions of (see types of); and governance, xiii, 56–71; history of, xiii, 1–20; implementation of, xiii, 7–10, 119–36; and institutions, 60–71; and interdependent science, 166–80; and policy, xiii, 64–6; and research, xiii, 36–55, 159; types of, xiii, 10–13
integrated water resource management, 226
integrated watershed management, 97–118; conceptual foundations of, 100–2
integration: across boundaries, 254; of civil and conventional science, 164–80; and communication, 125; and conflict management, 137–63; development and diversity in, 2–6; dimensions of, 41–8; drivers of, 38; of environmental, economic, and social goals, 204, 206, 216–17; and institutions, 45; of scientific knowledge and policy, 123–4; of stakeholder interests, 181–203; and sustainability assessment, 72–96; of theory and practice, 158–9; transboundary, 236–51
integration imperative, 36–41, 61
interactive mapping, 154–5
inter-agency: communication, 125; cooperation, 242, 246; tension, 126, 127, 128

Interagency Grizzly Bear Committee (IGBC), 143, 144
interdependent science, 166–80
International Association for Impact Assessment, 80
International Institute of Applied Systems Analysis, 105
International Joint Commission (IJC), 220
International Panel on Climate Change, 62
International Peace Park, 239, 241, 242
intra-disciplinary, 44

Jackson Lumber Company, 211–12
Jackson, M.C., 231
Jacobs, J., 57
Jacobs, K.L., 190
Jarita Mesa Lumber Company, 208
Jenkins, H.A., 27
Johannesburg Declaration, 37
Johannesburg (South Africa), 24
Johnson, B.L., 106
joint fact finding (JFF), xiv, 181–203; and Cancer Incidence Study, 192–9; and stakeholder interests, 194
Joint Strategic Plan for Management of Great Lakes Fisheries (GLFC), 220
Jones, C.O., 122
Jonkel, J.J., 153, 157
Jordan, T.G., 142–3
Josselson, R., 63

Kananaskis (Alberta), 236
Karl, H.A., 191
Kasworm, W., 145
Kay, J.J., 231
Kemp, W.M., 7
Kennett, S.A., 1, 246
Kimmins, J.P., 4
Kirk, J., 222
Knetsch, J., 45
Knight, R.L., 153
knowledge management, 220–35
Koch, M., 23–4
Kooiman, J., 225
Kopas, P., 125
Kranz, R., 106
Kreutzwiser, R., 228

Krutilla, J.V., 2, 99
Kuhn, T., 99

La Compañía Ocho logging company, 213–15
Land and Water Australia, xiii, 1, 48
Landres, P.B., 238
land-use: options, 40; planning, xii, 1, 3, 73, 77, 92; practice, 145; systems, 215
Lang, R., xii, 11, 12
Lant, C., xiii, 97–118
Las Comunidades, 214–15
Lasswell, H.D., 139
Latour, B., 167
Law of the Sea, 23
Lawrence, J., 104
Lee, K.N, 104–6
Lesser, E., 57
Linder, S.H., 122
Lindh, C.O., 208, 211
Linnaeus, 167
Liverpool Plains, 51, 105
Long, B., 240
Lord, G.R., 27
Lynch, K., 168

McCarthy, D., xiv, 220–35, 65
McConnaha, W.E., 104
McCormick, J., 22, 24
McCreary, S., 190
Macdonald Commission, 58
McElroy, M.W., 223–4
McGinnis, M.V., 104
MacIver, I., 29
McKay, 6, 81
McKearnan, S., 192
McLean, R.J., 168
McLellan, B.N., 143
McPhee, M.W., 126
Madera Forest Products Association (MFPA), 213–15
Madonna, K.J., 236
Maine, xiv, 192–9
Maine Cancer Registry, 194, 196
Maine Department of Environmental Protection (DEP), 193
Margerum, R.D., 4, 13, 122, 238, 239
Marin-Hernandez, A., 164

Martensson, M., 223, 224
Maslow's fixed hierarchy of individual needs, 56
Mason, D., 207
Mattson, D.J., 137, 139, 143
Mazmanian, D.A., 122, 123
Mexican spotted owl, 214–15
Michaels, S., xiv, 220–35
migrant workers, 168–80
Miistakis Institute, 241
Miller, F.J., 222
Mills, B., 62
Mitchell, B., xii, xiii, 1, 4, 6, 10, 11, 21–35, 122, 123, 124, 125, 226, 232
modelling, 2, 3, 49, 105, 112–14
Montana, xiv, 137–63
Montana Department of Fish, Wildlife, and Parks (FWP), 139, 141, 142, 144, 146, 147, 153, 156
Moote, M.A., 24, 26
Mortsch, L.D., 62
multi-agency review, 126, 127
multi-criteria analyses (MCA), 44, 110
multi-dimensional integrated approaches, 6
multi-disciplinary, xiv, 8, 36–55, 58
multiple methods, 158, 160
multiple use, xii, 2, 3, 5, 6, 25–6, 28, 30
Multiple-Use Sustained-Yield Act, 24–5
multi-scale analysis, 11, 254
multi-stakeholder, xiv, 11, 31, 45, 92
Munich Wildlife Society, 245
Munson, A., 23–4
Munton, R., 46
Murray, K., 137
Musiana, M., 237
Muth, R., 139

National Dryland Salinity Program (NDSP), 48, 105
National Environmental Policy Act, 25, 26, 77
National Forest Management Act (NFMA), 25, 212
National Research Council (NRC), 113
National Trails Act, 25
Natural Resources Conservation Service (NRCS), 156
negotiation, 12, 122, 131, 173, 184, 200

neighbour network, 156–7
Nelson, J.G., 30, 65
New Deal, 98, 207
New Mexico, xiv, 204–19
New Mexico Department of Game and Fish, 214
Newman, L., xiii, 56–71
New South Wales, 51
Nonaka, I., 222
non-timber forest product (NTFP), 164–80
non-timber forest product harvesters, 164–80
North American Conservation Conference, 22
Northern Oxford County Coalition (NOCC), xiv, 192–9
Northwest Planning Council, 104
Northwest Research and Harvesters Association (NRHA), 167–80
Norton, B.D., 4, 5
Nowicki, P., 242

O'Connor, D.R., 229
Ohio, 27, 226
Ohio Conservation Districts, 226
Olson, D.M., 243
Olympic National Forest, 170
Olympic Peninsula, xiv, 167–80
Olympic Peninsula Salal Sustainability and Management Research Project (*see* Salal Project)
Ontario, 220–35
Ontario Conservation Authorities, xiii, xiv, 24, 27–30, 221–35
Ontario Ministry of Natural Resources (MNR), 226, 228
organizational management, 159
Orna, E., 222
Oxford County (Maine), 181–203
Ozawa, C.P., 190; 191

Pacific Northwest, 154–80
Paehlke, R., 125
panarchy, xiii, 56, 62, 99–100
Paquet, P.C., 236, 237
Parker, D.C., 114
Parris, T.M., 239

participation, 252–3; of agencies, 128–9; community, 42; and integration, 38–55, 123; public, 26, 29, 30, 56, 58, 73–4, 76–7, 128–9, 131, 244
participatory processes, xiii, xiv, 3–5, 9–13, 26, 76–7, 155, 199
participatory research, 166–80
Pawson, E., 43
Pearse, P.H., 6
Pearson, G.A., 25
perceptions, attitudes, and values (PAV), 7, 9–10
permitting processes, 164–80; 184
Peters, B.G., 122
Petty, J.N., 100
Peyser, J., xiv, 181–203
Piatra Craiului National Park, 245
Pieniny International Landscape Park, 239
pillar approach, 83
Pinchot, G., 3, 24
Platt, J., 124
'Pluie,' 236–7
policy development, xiii, xiv, 10–11, 36–55, 56–71, 109–10, 132, 158, 202, 230
policy implementation, 119–36
policyscape, 140
'polluter pays' principle, 244
Pooler, F., 210, 216
post-audits, 105
Prato, T., 110
precautionary principle, 244
Pretty, J.N., 100
Price, R., xiii, 36–55
Primm, S.A., 137, 139, 144, 146
production possibilities frontier (PPF), 114
Promberger, C., 245
protected areas, 2
prototyping, 139, 159–60
Provincial Land and Resources Cluster of Ministries (PLRCM), 226
Prusak, L., 57
Public Health, US Department of, 194
Putman, R., 57

Quinn, M.S., xiv, 236–51

Ramadier, T., 61

Ransmeier, J.S., 27
Rayner, S., 124
Reason, P., 173, 174
Regier, H.A., 5
regional perspective/approach, 97, 104, 107, 240
renewable resources, 4, 27
research: collaborative/participatory, 36–55, 62, 173–4, 244; integrated, 36–55; interdependent, 164–80; interdisciplinary, 36–55; 62; multi-disciplinary, 36–55; policy-oriented, 43–4, 47; single sector/discipline, 39, 42; transdisciplinary, 60
resource and environmental management (REM), 181, 183, 252; approaches to, 3; areas of application and development, 2; changes in, xii; and economic benefits, 2; integrated approaches to, 125; and integration, xii–iv, 1–20, 21–35; language of, 12; manifestations of, xii; types of, xiii
resource management, 73, 109, 114; and economic benefits, 2
'responsive management,' 61
Rhodes, R.A.W., 224
Rice, D.S., 104
Rice, P.M., 104
Richardson, A.H., 27, 28
Rio Declaration on Environment and Development, 23, 24, 37
Rio Summit, 23
river basin management, 99, 109
river valley development, 27
River Valley Health Communities Coalition, 199
Robinson, J.B., 58, 60, 62
Rocheleau, B., 223
Rocky Mountain Front, 137–63
Rocky Mountains, 236, 238, 240–2
Romanian State Forest Administration, 245
Roosevelt, President Theodore, 22, 207
Rossi, P., 122
Rothman, D., 58, 62
Rouse, W.B., 221, 222, 223
Ruhl, J.B., 109
Rumford (Maine), xiv, 192, 196, 198

Sabatier, P.A., 122, 123
Safe Water Drinking Act, 30
salal, xiv, 164–80; brush, 170, 174; heavy intensity harvest of, 169–70, 175; light intensity harvest of, 170, 175
salal harvesters, 164–80; and ecological knowledge, 169–80; and permits to harvest, 168–80
Salal Project, 166–80; and forest management, 175; and interdependent science, 164–80; and policy, 172, 176
Salwasser, H., 26
Sanderson, E.W., 238
Schallau, C.H., 207
Schoenberger, E., 42
Scott, D., 206, 207
Seitz, W., 105
self-organizing systems, 4–5
sensitivity analysis, 190
Servheen, C., 137, 144
Shackley, S., 58
Shah, T., 109
Shearer, K., 221, 222, 223
Shrubsole, D., xiii, 21–35, 226
Sick, D., 239
Sierra Club, 216
silos, 58–60
Simard, A.J., 221, 222
Sime, J., 60
Skogstad, G., 125
Slocombe, D.S., xiii, 1–20, 119, 220, 232
Smith, A.F., 29
social trap, 124
Society of American Foresters, 3
socio-ecological systems, 239
solitudes, 58, 60
specialized generalists/generalized specialists, 8
Steiner, M., 24
Stinson, B.L., 190
Stockholm Conference, 29, 58
Stockholm Declaration on Human Evolution, 22, 24
Stockholm (Sweden), 22
stovepipes, 58, 60
Sullivan, R., 23–4
Surveillance, Epidemiology, and End Results Program (SEER), 197

Susskind, L., xiv, 123, 181–203
sustainability, 38, 41, 42, 46, 53, 72–96, 105–6, 110, 175, 204, 206, 215, 217, 243, 255; concepts of, 82–5; of ecosystem, 106; emergence of, 80; and equity, 216; requirements, 83–5; of resources, 4, 58, 168, 170; of wildlife, 241
sustainability assessment, xiii, 72–96; future of, 91–3; implementation of, 88–91; and integration, 72–96; and intertwined considerations, 91; processes of, 85–7; requirements of, 84; rise of, 81–2
sustainable development, xiii, 23–4, 37, 45, 59, 60, 63, 81, 244; and institutions, 61–9; multi-scaled approaches to, 57
sustainable: forestry, 214; management practices, 175; use, 244
sustained yield, xiv, 204–19
Sustained Yield Forest Management Act (SYFMA), 207, 212, 217
systems approach/perspective: orientation of, 53; theories of, xiii, 1, 2, 14, 32, 98, 231, 232

Taft, President William H., 22
Takeuchi, H., 222
Taylor, D., 139
Tennessee Valley Authority, 27, 99, 226
Teton River Watershed Group, 146–52
Thomson, K., 23–4, 81
Thorsell, J., 239
Tilman, D., 100
timber management practices, 174
Tinker, J., 60
Torgerson, D., 125
Total Maximum Daily Load (TMDL), 97
Trade-offs, 88–91, 114
transboundary national resource management (TBNRM), xiv, 236–51
transboundary: protected areas, xiv, 239; regions, 236–51; wildlife management, 236–51
transdisciplinary, 59, 60, 63
triple bottom line accounting, 44
Trist, E.L., 57, 59, 60
Turnock, D., 243

United Nations Conference on the Environment and Development (UNCED), 22, 23–4, 58
United Nations Conference on the Human Environment, 22
United Nations Environmental Programme (UNEP), 23, 245
urban: design, 73; development, 130; sustainability, 81

Vallecitos Federal Sustained Yield Unit (VFSYU), 204–19
Vallecitos Forest, xiv, 204–19
Vallecitos Lumber Company, 211
Vancouver (British Columbia), 81, 119–36
van der Linde, H., 239
van der Wansen, M., xiv, 181–203
Van Horn, C.E., 122
van Kerkhoff, L., 1, 8
Van Meter, D.S., 122
Veale, B., 30
Victoria (Australia), 105
Vienna (Austria), 105, 244
Vigil, G., 213
Volkman, J.M., 104
Voloscuk, I., 243

Walkerton (Ontario), 30, 229
Wall, L.A., 211
Wallace, R.L., 140
Walters, C., 104, 105
Walther, P., 5, 10, 11, 122, 124, 125
Wartenberg, D., 195–9
Washington Department of Natural Resources, 175
Washington (District of Columbia), 22, 211
Washington State, xiv
water quality, 97, 126, 130, 146, 216, 220–30
water resource development, 97
Water Resource Information Project (WRIP), xiv, 226–9
water resource management, 220–35
watershed management, xiii, xiv, 1–3, 8, 27–9, 97–118, 190, 221, 252
Waterton Lakes National Park, 236, 239, 241
Weale, A., 97, 122